APPLIED RESEARCH IN
FIELD CROP PATHOLOGY
FOR INDIANA, 2023

APPLIED RESEARCH IN FIELD CROP PATHOLOGY FOR INDIANA, 2023

DARCY E. P. TELENKO
AND SUJOUNG SHIM

PURDUE UNIVERSITY PRESS
WEST LAFAYETTE, INDIANA

Cataloging-in-Publication Data on file at the Library of Congress.

978-1-62671-249-2 (paperback)

978-1-62671-250-8 (epdf)

CONTENTS

ACKNOWLEDGMENTS

This report is a summary of applied field crop pathology research trials conducted in 2023 under the direction of the Field Crop Pathology program in the Department of Botany and Plant Pathology at Purdue University. The authors wish to thank the Purdue Agronomy Research and Education Center, the Purdue Agricultural Centers, and the many cooperators and contributors who provided the resources needed to support the applied field crop pathology research program in Indiana.

Special recognition is extended to Stephen Brand and Su Shim for technical skills in managing field trials and data organization and processing and for helping to prepare this report; Camila Rocca da Silva, Morgan Goodnight, Monica Mizuno, Ivis Miranda, and Nileshwari Yewle, graduate students and visiting scholars who assisted with field trial data collection and analysis; Emily Duncan, Cora Reynolds, Emilia Meyers, and Joel Will, undergraduate student interns who assisted with field trial data collection and scouting; Dr. Tom Creswell, Dr. John Bonkowski, and Tina Garwood with the Purdue Plant Pest Diagnostic Laboratory for assistance in pathogen surveys and diagnosis. Collectively, the contributions of colleagues, professionals, students, and growers were responsible for a highly successful and productive program to evaluate products and practices for disease management in field crops.

The authors would also like to thank the following for their support in 2023: Adama; Albaugh; AMVAC; Ascribe Bioscience; Bayer Crop Science; BASF; Corteva Agriscience FMC Agricultural Solution; Germains; Gowan; the Helena, Indiana, Corn Marketing Council; the Indiana Soybean Alliance; Koppert; the North Central Soybean Research Program NC SARE Project # LNC20–443; Oro Agri/Rovensa; ProFarm/Maron Bio; Pioneer; Purdue University; Sipcam Agro, Syngenta the USDA NIFA Hatch Project #1019253; the USDA NIFA CARE Project #2021–09839; USWBSI-NFO; UPL; Valent and VM Agritech.

SUMMARY OF 2023 FIELD CROP DISEASE SEASON

CORN

In 2023 there was moderate disease on corn in Indiana across the state. Details of major issues are listed below. Gray leaf spot, northern corn leaf blight, northern corn leaf spot, and southern rust were found in pockets. Tar spot and southern rust were two diseases that were closely monitored this season. In addition, at the conclusion of the season there were numerous reports about issues with mycotoxins in corn. For more information about mycotoxins and resources available on management, storage, and testing, see https://cropprotectionnetwork.org/publications/mycotoxin-faqs.

Tar spot: Tar spot of corn was a concern due to previous epidemics. In 2023, moderate levels of tar spot occurred in northern Indiana and in pockets in other areas of the state. The environmental conditions are key in determining field risk year to year, as leaf wetness plays an important role in tar spot disease development. The fifth year of tar spot–directed research has been completed here in Indiana. As a cautionary note, it is still important to have multiple years of data for verification, but the initial results do serve as a good starting point for making future management decisions.

We continue to scout for tar spot across the state. One new county was confirmed with tar spot in 2023, making 87 counties total in Indiana to date (Figure 1). It is important to document tar spot movement in the state so that when favorable conditions arise, increased tar spot disease risk can be more accurately assessed across the remainder of the state.

Southern corn rust: Southern corn rust was first found in Indiana in the 2023 season on August 25, and by the end of the season a total of seven counties were confirmed to have the disease present (Figure 2a). Southern rust pustules generally tend to occur on the upper surface of the leaf and produce chlorotic symptoms on the underside of the leaf (Figure 2b). These pustules rupture the leaf surface and are orange to tan in color. They are circular to oval in shape. Common rust was also widespread, and both diseases could be present on a leaf and easily mistaken for each other. It is important to send a sample to the Purdue Plant Pest Diagnostic Lab for confirmation if southern rust is suspected. There is an increased risk for yield impact if southern rust is identified early in the season.

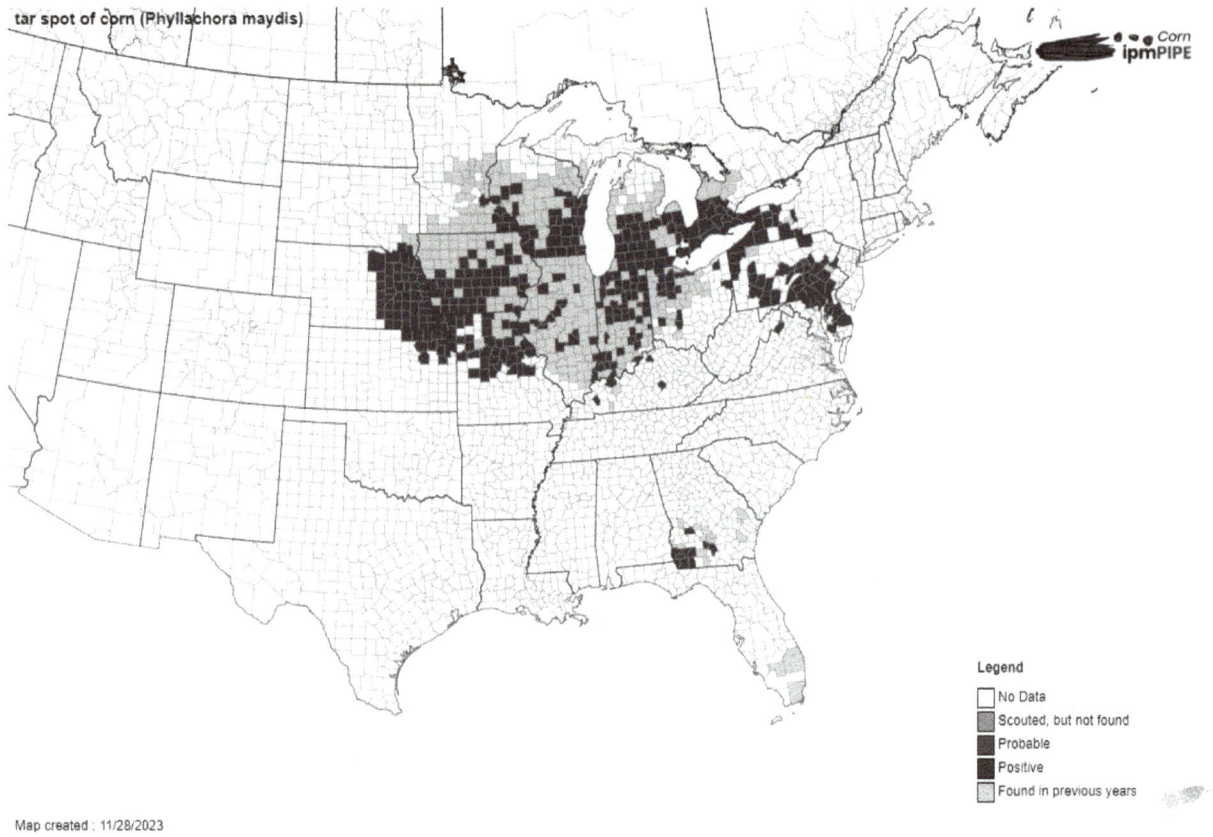

FIGURE 1. 2023 tar spot tracking across the United States and Canada. Dark gray indicates that a positive sample was collected from that county during the 2023 season, and lighter gray indicates that tar spot has been found previously. Image source: IPM Pipe, https://corn.ipmpipe.org/tarspot/ on 11/28/2023.

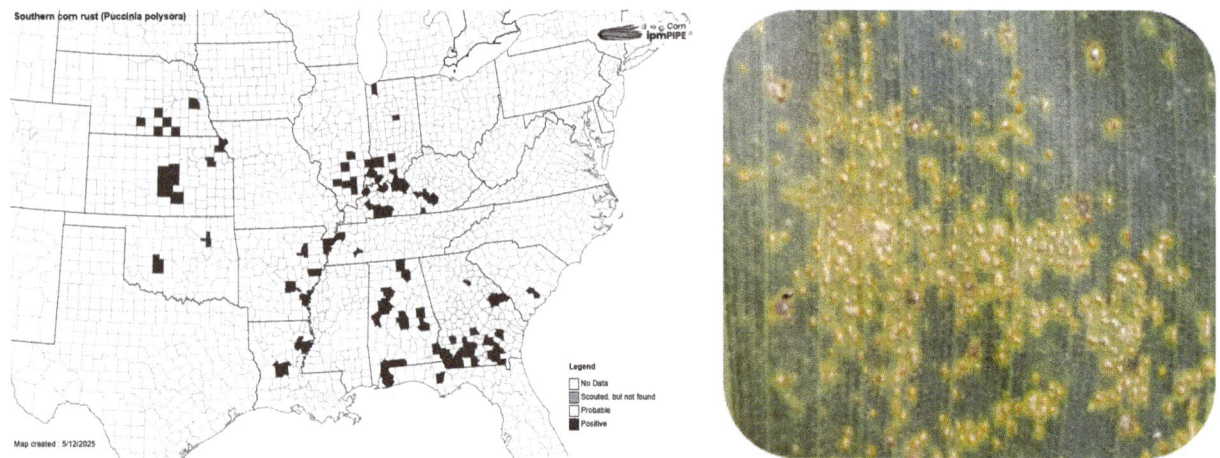

FIGURE 2. Southern corn rust map (*left*) of confirmed (dark gray) counties that had southern corn rust in 2023 and a corn leaf with southern rust infection (*right*). Photos credit: D. Telenko, Map source: Crop Protection Network, https://corn.ipmpipe.org/southerncornrust/.

SOYBEAN

Diseases in soybeans remained relatively low throughout the season for much of the state except for areas with white mold. These areas experienced high levels of disease this season due to optimum conditions for disease infection and development. Our research sites and sentinel plots across the state saw low levels of frogeye leaf spot, Cercospora leaf blight, and Septoria brown spot. There were also pockets where sudden death syndrome caused issues in fields, and we had new reports of red crown rot, which has now been confirmed in four counties (Adams, Decatur, Rush, and Spencer).

WHEAT

Fusarium head blight (FHB), or scab, is one of the most impactful diseases of wheat and among the most challenging to prevent. In addition, FHB infection can cause the production of a mycotoxin called deoxynivalenol (DON, or vomitoxin). The conditions in 2023 were not conducive to FHB development. Our research sites in both West Lafayette and Vincennes had extremely low levels of FHB develop in our nontreated susceptible cultivar checks and little to no DON detected in our grain. FHB management requires an integrated approach, including the selection of cultivars with moderate resistance and timely fungicide application at flowering. Very few other diseases were observed in our wheat trials.

AGRONOMY CENTER FOR RESEARCH AND EDUCATION (ACRE)

EVALUATION OF FUNGICIDES FOR FOLIAR DISEASES IN CORN IN CENTRAL INDIANA, 2023 (COR23-01.ACRE)

E. A. Duncan, S. Shim, and D. E. P. Telenko, Department of Botany and Plant Pathology, Purdue University West Lafayette, IN 47907-2054

CORN (ZEA MAYS P0574AM)

Tar spot, *Phyllachora maydis*
Gray leaf spot, *Cercospora zeae-maydis*
Northern corn leaf blight, *Exserohilum turcicum*

A trial was established at the Purdue Agronomy Center for Research and Education (ACRE) in Tippecanoe County, Indiana. The trial was a randomized complete block design with four replications. Plots were 10 feet wide and 30 feet long and consisted of four rows, and the two center rows were used for evaluation. The field was no-till and had been continuous corn for over five years. Standard practices for nonirrigated grain corn production in Indiana were followed. Corn hybrid P0574AM was planted in a no-till field in 30-inch row spacing at a rate of 2 seeds/foot on May 11. All foliar fungicide applications were applied at 15 gal/acre and at 40 psi using a Lee self-propelled sprayer equipped with a 10-foot boom, fitted with six TJ-VS 8002 nozzles spaced 20 inches apart. Fungicides were applied on July 31 at tassel/silk (VT/R1) growth stage. Disease ratings were assessed on September 21 at physiological maturity (R6) growth stage. Tar spot, gray leaf spot (GLS), and northern corn leaf blight (NCLB) severity was visually assessed as a percentage (0–100%) of symptomatic leaf area on ear leaf, and five plants were assessed per plot and averaged before analysis. Percent of canopy greenness was rated by visually assessing the percentage (0–100%) of the whole plot for crop canopy that remained green at physiological maturity (R6) growth stage on September 22. The two center rows of each plot were harvested on October 17, and yields were adjusted to 15.5% moisture. All data were analyzed in SAS 9.4 (SAS Institute, Cary, NC). A generalized linear mixed model analysis of variance was performed using PROC

GLIMMIX. Values are least squares means, and values with different letters are significantly different based on the least squares difference test (α = 0.05).

In 2023, weather conditions were moderately favorable for disease development. Tar spot, GLS, and NCLB were the most prominent diseases in the trial. All fungicide treatments reduced tar spot severity and GLS over the nontreated control (Table 1). All fungicides treatments reduced NCLB severity over the nontreated control except Delaro Complete. Veltyma, Delaro Complete, Adastrio, Miravis Neo, Proline, and Quadris significantly increased canopy greenness over the nontreated control on September 22. No significant differences were detected for yield of corn.

TABLE 1. *Effect of Treatment on Foliar Disease Severity, Canopy Greenness, and Yield of Corn*

TREATMENT AND RATE/ACRE[z]	TAR SPOT[y] %	GLS[y] %	NCLB[y] %	CANOPY GREEN[x] %	YIELD[w] BU/ACRE
Nontreated control	0.43 a	1.4 a	6.9 a	70.0 c	198.3
Veltyma 3.34 S 7.0 fl oz	0.05 b	0.1 c	1.0 bcd	78.8 ab	203.3
Delaro Complete 458 SC 8.0 fl oz	0.08 b	0.2 c	3.9 ab	82.5 a	200.5
Aproach Prima 2.34 SC 6.8 fl oz	0.13 b	0.4 bc	2.5 bcd	70.0 c	201.7
Adastrio 4.0 SC 8.0 fl oz	0.08 b	0.2 c	0.4 d	81.3 ab	200.6
Miravis Neo 2.5 SE 13.7 fl oz	0.10 b	0.2 c	0.9 bcd	73.8 bc	199.0
Trivapro 2.21 SE 13.7 fl oz	0.18 b	0.1 c	0.6 cd	77.5 abc	215.9
Headline AMP 1.68 SC 10 fl oz	0.13 b	0.4 bc	3.6 bc	73.8 bc	198.8
Proline 480 SC 5.7 fl oz	0.08 b	0.2 c	1.4 bcd	78.8 ab	203.1
Quadris 2.08 SC 6.0 fl oz	0.08 b	0.7 b	3.3 bcd	81.3 ab	207.6
Tilt 3.6 EC 4.0 fl oz	0.18 b	0.2 c	2.0 bcd	76.3 abc	203.4
P-value[v]	0.0040	0.0001	0.0073	0.0264	0.3524

[z] Fungicide treatments were applied on July 31 at tassel/silk (VT/R1) growth stage.

[y] Foliar disease severity was visually assessed as a percentage (0–100%) of symptomatic leaf area on ear leaf, with five plants assessed per plot and averaged before analysis on September 21 at physiological maturity (R6) growth stage. Tar spot stromata severity rated. GLS = gray leaf spot, NCLB = northern corn leaf blight.

[x] Canopy greenness was visually assessed as a percentage (0–100%) of crop canopy green on September 22.

[w] Yields were adjusted to 15.5% moisture and harvested on October 17.

[v] All data were analyzed in SAS 9.4 (SAS Institute, Cary, NC). A generalized linear mixed model analysis of variance was performed using PROC GLIMMIX. Values are least squares means, and values with different letters are significantly different based on the least squares difference test (α = 0.05).

EVALUATION OF TILLAGE, HYBRID, AND FUNGICIDE EFFICACY FOR DISEASES IN CORN IN CENTRAL INDIANA, 2023 (COR23-06.ACRE)

S. Shim and D. E. P. Telenko, Department of Botany and Plant Pathology, Purdue University
West Lafayette, IN 47907-2054

CORN (*ZEA MAYS* W2585SSRIB, P0589AMXT)

Tar spot, *Phyllachora maydis*
Gray leaf spot, *Cercospora zeae-maydis*
Northern corn leaf blight, *Exserohilum turcicum*

A trial was established at the Purdue Agronomy Center for Research and Education (ACRE) in Tippecanoe County, Indiana. The trial was a randomized complete block design with six replications. Plots were 10 feet wide and 30 feet long and consisted of four rows, and the two center rows were used for evaluation. The previous crop was corn. Standard practices for nonirrigated grain corn production in Indiana were followed. A two-tillage block of no-till and standard tillage was the main effect. Corn hybrid W2585SSRIB (tar spot susceptible) and P0589AMXT (tar spot moderate resistant) were planted in 30-inch row spacing at a rate of 2 seeds/foot on May 5. Veltyma fungicide was applied at blister (R2) growth stage. Disease severity was rated by visually assessing the percentage of symptomatic leaf area per ear leaf on 10 plants in each plot and averaged before analysis. Tar spot severity was visually assessed on September 26 at maturity (R6) growth stage. Gray leaf spot (GLS) and northern corn leaf blight (NCLB) severity were visually assessed as a percentage (0–100%) of symptomatic leaf area on ear leaf on September 12 at dent/maturity (R5/R6). The two center rows of each plot were harvested on October 18, and yields were adjusted to 15.5% moisture All disease and yield data were analyzed in SAS 9.4 (SAS Institute, Cary, NC). A generalized linear mixed model analysis of variance was performed using PROC GLIMMIX. Values are least squares means, and values with different letters are significantly different based on the least squares difference test ($\alpha = 0.05$).

In 2023, weather conditions were moderately favorable for disease. GLS and NCLB were the most prominent diseases in the trial and reached low severity. When no significant interactions between tillage, hybrid, and fungicide treatments were detected, the main effects were separated (Table 2). For tar spot and yield there was a significant interaction between tillage and hybrid, so means were separated for that effect. There was no significant effect on Veltyma application for tar spot severity compared to the nontreated control. Tar spot was significantly reduced by W2585SSRIB planted in a tilled low residue or when P0589AMXT was planted as compared to W2585SSRIB in no-till. GLS severity was significantly reduced with tillage (low residue) as compared to no-till, with Veltyma fungicide application as compared to the nontreated control and in the hybrid P0589AMXT as compared to W2585SSRIB. There were no significant differences between hybrids for NCLB. NCLB was lower in the tilled plot verses the no-till plot and was significantly reduced by Veltyma over nontreated. Grain harvest moisture was higher in the no-till plot, on hybrid W2585SSRIB, and with fungicide treatment. Test weight was significantly higher under tillage and with the hybrid P0589AMXT. Both hybrids planted into tilled plots increased yield, while W2585SSRIB resulted in the highest grain yield in the low-residue plots.

TABLE 2. *Effect of Tillage, Hybrid, and Fungicide for Foliar Disease Risk in Corn and Yield of Corn*

TILLAGE, HYBRID, TREATMENT, AND TIMING[z]	TAR SPOT[y] %		GLS[y] %	NCLB[y] %		HARVEST MOISTURE %	TEST WEIGHT LB/BU	YIELD[x] BU/ACRE	
No tillage (high residue)	0.3		0.8 b	2.9		19.3 a	55.2 b	184.7	
Yes tillage (low residue)	0.0		1.7 a	2.1		17.0 b	56.3 a	208.7	
	No-till	Tilled						No-till	Tilled
Tar spot susceptible	0.5 a	0.0 b	0.8 b	2.7		18.6 a	54.9 b	190.6 b	220.9 a
Tar spot moderate resistant	0.1 b	0.0 b	1.7 a	2.3		17.7 b	56.6 a	178.7 c	195.4 b
				No-till	Tilled				
Nontreated control	0.2		1.6 a	5.4 a	3.1 b	17.8 b	55.9	194.4	
Veltyma 3.34 SC 7.0 fl oz Acre	0.1		0.9 b	0.4 c	1.2 c	18.5 a	55.5	198.4	
P-value *tillage*[w]	*0.0001*		**0.0001**	*0.1856*		**0.0001**	**0.0102**	*0.0001*	
P-value *hybrid*	*0.0003*		**0.0005**	*0.4334*		**0.0075**	**0.0007**	*0.0001*	
P-value *fungicide*	*0.1888*		**0.0057**	*0.0001*		**0.0342**	*0.3518*	*0.2068*	
P-value *tillage*hybrid*	**0.0005**		*0.1823*	*0.3394*		*0.3554*	*0.9087*	**0.0376**	
P-value *tillage*fungicide*	*0.1406*		*0.1502*	**0.0079**		*0.1018*	*0.4026*	*0.7028*	
P-value *hybrid*fungicide*	*0.1888*		*0.3641*	*0.2369*		*0.5128*	*0.6647*	*0.9070*	
P-value *tillage*hybrid* fungicide*	*0.1406*		*0.2330*	*0.6625*		*0.3989*	*0.4132*	*0.6177*	

[z] Veltyma application was applied at 15 gal/acre and at 40 psi using a Lee self-propelled sprayer equipped with a 10-foot boom, fitted with six TJ-VS 8002 nozzles spaced 20 inches apart. Veltyma was applied on August 2 at blister (R2) growth stage.

[y] Foliar disease severity visually was assessed as a percentage (0–100%) of symptomatic leaf area on ear leaf, with 10 plants assessed per plot and averaged before analysis on September 26 at physiological maturity (R6) for tar spot and on September 12 at dent/maturity (R5/R6) for GLS and NCLB. GLS = gray leaf spot, NCLB = northern corn leaf blight.

[x] Yields were adjusted to 15.5% moisture and harvested on October 18.

[w] All data were analyzed in SAS 9.4 (SAS Institute, Cary, NC). A generalized linear mixed model analysis of variance was performed using PROC GLIMMIX. Values are least squares means, and values with different letters are significantly different based on the least squares difference test (α = 0.05).

EVALUATION OF FUNGICIDES FOR FOLIAR DISEASES IN CORN IN CENTRAL INDIANA, 2023 (COR23-15.ACRE

E. A. Duncan, S. Shim, and D. E. P. Telenko, Department of Botany and Plant Pathology, Purdue University West Lafayette, IN 47907-2054

CORN (*ZEA MAYS* P0574AM)

Tar spot, *Phyllachora maydis*
Gray leaf spot, *Cercospora zeae-maydis*
Northern corn leaf blight, *Exserohilum turcicum*

A trial was established at the Purdue Agronomy Center for Research and Education (ACRE) in Tippecanoe County, Indiana. The trial was a randomized complete block design with four replications. Plots were 10 feet wide and 30 feet long and consisted of four rows, and the two center rows were used for evaluation. The field was no-till and had been continuous corn for over five years. Standard practices for grain corn production in Indiana were followed. Corn hybrid P0574AM was planted in 30-inch row spacing at a rate of 2 seeds/foot on May 11. Fungicide treatments were made on August 1 at blister (R2) growth stage. All foliar fungicide applications were applied at 15 gal/acre and 40 psi using a Lee self-propelled sprayer equipped with a 10-foot boom, fitted with six TJ-VS 8002 nozzles spaced 20 inches apart. Disease ratings were assessed on September 18 at dent (R5) growth stage. Tar spot, gray leaf spot (GLS), and northern corn leaf blight (NCLB) severity were visually assessed as a percentage (0–100%) of symptomatic leaf area on ear leaf, and five plants were assessed per plot and averaged before analysis. The two center rows of each plot were harvested on October 17, and yields were adjusted to 15.5% moisture. All data were analyzed in SAS 9.4 (SAS Institute, Cary, NC). A generalized linear mixed model analysis of variance was performed using PROC GLIMMIX. Values are least squares means, and values with different letters are significantly different based on the least squares difference test (α = 0.05).

In 2023, weather conditions were moderately favorable for disease. GLS and NCLB were the most prominent diseases in the trial. No significant differences between treatments and the nontreated control were detected for tar spot and GLS (Table 3). All foliar fungicides significantly reduced NCLB symptoms over the nontreated control. All fungicides reduced lodging over the nontreated control except Soratel 5.0 fl oz + Aproach 4.75 fl oz, Soratel at 2.5 fl oz and 5.0 fl oz, and Aproach at 3.5 and 6.75 fl oz. No significant differences between treatments and the nontreated control were detected for test weight and yield of corn.

TABLE 3. *Effect of Treatments on Foliar Disease Severity, Lodging, and Yield of Corn*

TREATMENT AND RATE/ACRE[z]	TAR SPOT[y] %	GLS[y] %	NCLB[y] %	LODGING[x] %	TEST WEIGHT LB/BU	YIELD[w] BU/ACRE
Nontreated control	0.00	2.2	8.4 a	15.0 a	55.9	186.4
Soratel 250 EC 2.5 fl oz + Approach 2.08 SC 3.5 fl oz	0.00	1.7	1.4 b	0.0 d	53.8	195.7
Soratel 250 EC 3.5 fl oz + Approach 2.08 SC 4.75 fl oz	0.00	1.0	0.9 b	2.5 cd	55.3	189.4
Soratel 250 EC 5.0 fl oz + Approach 2.08 SC 6.75 fl oz	0.03	1.4	3.0 b	7.5 a-d	54.9	195.9
Stratego 2.08 SC 4.0 fl oz	0.00	0.8	2.0 b	5.0 bcd	54.1	196.7
Stratego 2.08 SC 5.0 fl oz	0.00	0.7	3.6 b	2.5 cd	59.0	195.9
Soratel 250 EC 2.5 fl oz	0.00	1.2	2.8 b	10.0 abc	54.4	190.2
Soratel 250 EC 5.0 fl oz	0.03	1.3	3.0 b	7.5 a-d	54.1	191.9
Aproach 2.08 SC 3.5 fl oz	0.00	1.8	1.9 b	12.5 ab	54.0	191.9
Aproach 2.08 SC 6.75 fl oz	0.00	1.1	2.1 b	12.5 ab	54.6	186.1
Maxentis 2.92 SC 8.0 fl oz	0.00	1.1	2.1 b	2.5 cd	54.6	192.9
Quadris 2.08 SC 3.5 fl oz	0.03	1.6	1.1 b	0.0 d	54.7	192.5
Quadris 2.08 SC 6.75 fl oz	0.00	1.7	2.2 b	2.5 cd	54.5	196.1
Flint Extra 4.08 SC 1.8 fl oz	0.00	0.5	3.3 b	0.0 d	54.4	188.6
Flint Extra 4.05 SC 3.45 fl oz	0.00	0.6	2.1 b	0.0 d	53.6	187.7
Aproach Prima 2.34 SC 6.4 fl oz	0.00	1.1	1.7 b	0.0 d	55.2	188.9
Veltyma 3.34 SC 7.0 fl oz	0.00	1.5	0.8 b	4.2 bcd	54.9	195.2
P-value[v]	*0.6066*	*0.2931*	*0.0091*	*0.0132*	*0.1390*	*0.7008*

[z] Fungicide treatments were applied on August 1 at blister (R2) growth stage.

[y] Foliar disease severity was visually assessed as a percentage (0–100%) of symptomatic leaf area on ear leaf, with five plants assessed per plot and averaged before analysis on September 18 at dent (R5) growth stage. GLS = gray leaf spot, NCLB = northern corn leaf blight.

[x] Lodging was assessed as a percentage of lodged stalks when pushed from shoulder height to at 45-degree from vertical on September 18.

[w] Yields were adjusted to 15.5% moisture and harvested on October 17.

[v] All data were analyzed in SAS 9.4 (SAS Institute, Cary, NC). A generalized linear mixed model analysis of variance was performed using PROC GLIMMIX. Values are least squares means, and values with different letters are significantly different based on the least squares difference test (α = 0.05).

EVALUATION OF FUNGICIDES AND APPLICATIONS FOR FOLIAR DISEASES IN CORN IN CENTRAL INDIANA, 2023 (COR23-25.ACRE)

E. A. Duncan, S. Shim, and D. E. P. Telenko, Department of Botany and Plant Pathology, Purdue University West Lafayette, IN 47907-2054

CORN (ZEA MAYS P0574AM)

Tar spot, *Phyllachora maydis*
Gray leaf spot, *Cercospora zeae-maydis*
Northern corn leaf blight, *Exserohilum turcicum*

A trial was established at the Purdue Agronomy Center for Research and Education (ACRE) in Tippecanoe County, Indiana. The experiment was a randomized complete block design with four replications. Plots were 10 feet wide and 30 feet long and consisted of four rows, and the two center rows were used for evaluation. The field was no-till and had been continuous corn for over five years. Standard practices for nonirrigated grain corn production in Indiana were followed. Corn was planted in 30-inch row spacing at a rate of 2 seeds/foot on May 5. Xyway 2x2 applications of 10 gal/acre were applied at planting. All foliar fungicide applications were applied at 15 gallon/acre and at 40 psi using a Lee self-propelled sprayer equipped with a 10-foot boom, fitted with six TJ-VS 8002 nozzles spaced 20 inches apart. Fungicides were applied on August 1, August 18, and August 25 at blister (R2), early dough (R4), and late dough (R4) growth stages, respectively. Disease ratings were assessed on September 14 at dent (R5) growth stage. Tar spot, gray leaf spot (GLS), and northern corn leaf blight (NCLB) severity were visually assessed as a percentage (0–100%) of symptomatic leaf area on the ear leaf, and five plants were assessed per plot and averaged before analysis. The two center rows of each plot were harvested on October 18, and yields were adjusted to 15.5% moisture. All data were analyzed in SAS 9.4 (SAS Institute, Cary, NC). A generalized linear mixed model analysis of variance was performed using PROC GLIMMIX. Values are least squares means, and values with different letters are significantly different based on least the squares difference test (α = 0.05).

In 2023, weather conditions were unfavorable for disease development. Tar spot, GLS, and NCLB were present in the trial but only reached low levels. Xyway followed by Adastrio 7.0 fl oz at late R4, Adastrio 8.0 fl oz at early R4, and Adastrio 8.0 fl oz at late R4 significantly reduced tar spot over the nontreated control (Table 4). All treatments reduced GLS over the nontreated control on September 14 except Xyway LFR 9.5 fl oz and Xyway LFR 15.2 fl oz. No significant differences between fungicide treatments and the nontreated control were found for NCLB severity, harvest moisture, test weight, and yield of corn.

TABLE 4. *Effect of Treatment on Foliar Disease Severity and Yield of Corn*

TREATMENT, RATE/ACRE, AND TIMING[z]	TAR SPOT[y] %	GLS[y] %	NCLB[y] %	HARVEST MOISTURE %	TEST WEIGHT LB/BU	YIELD[x] BU/ACRE
Nontreated control	0.13 ab	0.9 a	4.4	19.5	55.4	175.5
Xyway LFR 1.92 SC 9.5 fl oz 2x2	0.11 ab	1.1 a	2.4	18.2	56.0	190.3
Xyway LFR 1.92 SC 15.2 fl oz 2x2	0.08 abc	0.7 ab	2.4	18.8	54.9	187.4
Xyway LFR 1.92 SC 9.5 fl oz 2x2 fb Adastrio 4.0 SC 9.5 fl oz at R2	0.16 a	0.3 bc	0.6	19.5	55.0	190.3
Xyway LFR 1.92 SC 9.5 fl oz 2x2 fb Adastrio 4.0 SC 7.0 fl oz at R2	0.05 bc	0.1 c	0.7	19.0	56.1	199.9
Xyway LFR 1.92 SC 9.5 fl oz 2x2 fb Adastrio 4.0 SC 7.0 fl oz early at R4	0.04 bc	0.3 bc	2.0	19.5	55.7	182.7
Xyway LFR 1.92 SC 9.5 fl oz 2x2 fb Adastrio 4.0 SC 7.0 fl oz late at R4	0.00 c	0.4 bc	0.6	19.0	56.0	184.1
Adastrio 4.0 SC 8.0 fl oz at R2	0.06 bc	0.1 c	0.2	18.8	55.4	198.4
Adastrio 4.0 SC 8.0 fl oz at R2	0.09 abc	0.3 bc	0.3	18.8	55.7	190.4
Adastrio 4.0 SC 8.0 fl oz early at R4	0.02 c	0.3 bc	0.0	19.3	55.9	189.9
Adastrio 4.0 SC 8.0 fl oz late at R4	0.02 c	0.4 bc	1.2	18.9	55.3	195.2
P-value[w]	*0.0287*	*0.0006*	*0.1302*	*0.8650*	*0.5175*	*0.2295*

[z] Xyway 2x2 applications of 10 gal/acre were applied at planting. Fungicides were applied on August 1, August 18, and August 25 at blister (R2), early dough (R4), and late dough (R4) growth stages, respectively. All foliar fungicide applications were applied at 15 gal/acre. fb = followed by.

[y] Foliar disease severity was visually assessed as a percentage (0–100%) of symptomatic leaf area on ear leaf, with five plants assessed per plot and averaged before analysis on September 14 at dent (R5) growth stage. GLS = gray leaf spot, NCLB = northern corn leaf blight.

[x] Yields were adjusted to 15.5% moisture and harvested on October 18.

[w] All data were analyzed in SAS 9.4 (SAS Institute, Cary, NC). A generalized linear mixed model analysis of variance was performed using PROC GLIMMIX. Values are least squares means, and values with different letters are significantly different based on the least squares difference test (α = 0.05).

VALUATION OF IN-FURROW TREATMENTS FOR *FUSARIUM GRAMINEARUM* IN CORN IN CENTRAL INDIANA, 2023 (COR23-40.ACRE)

E. A. Duncan, S. Shim, and D. E. P. Telenko, Department of Botany and Plant Pathology, Purdue University West Lafayette, IN 47907-2054

CORN (*ZEA MAYS*)

Seedling disease, *Fusarium graminearum*

A trial was established at the Purdue Agronomy Center for Research and Education (ACRE) in Tippecanoe County, Indiana. The experiment was a randomized complete block design with four replications. Plots were 10 feet wide and 30 feet long and consisted of four rows, and the two center rows were used for evaluation. The field was no-till and had been continuous corn for over five years. Standard practices for nonirrigated corn production in Indiana were followed. Corn was planted in 30-inch rows spacing at a rate of two seeds/foot on May 5. Inoculum of *Fusarium graminearum* was applied within the seedbed at 1.25 g/foot at planting. In-furrow applications were applied at planting at 10 gal/acre. The two central rows of each plot were used for the stand counts on May 24, May 30, and June 28 at first trifoliolate (V1), second trifoliolate (V2), and fourth trifoliolate (V4) growth stages, respectively. Plant phytotoxicity (yellowing) was visually rated as a percentage (0–100%) per plot on May 30. The two center rows of each plot were harvested on October 18, and yields were adjusted to 15.5% moisture. All data were analyzed in SAS 9.4 (SAS Institute, Cary, NC). A generalized linear mixed model analysis of variance was performed using PROC GLIMMIX. Values are least squares means, and values with different letters are significantly different based on the least squares difference test ($\alpha = 0.05$).

In 2023 cool and wet conditions occurred after planting, which encouraged soil-borne disease to develop. No significant differences were detected between treatments for stand and phytotoxicity (Table 5). There were no significant differences between treatments and nontreated control for harvest moisture, test weight, and yield of corn.

TABLE 5. *Effect of In-Furrow Treatments on Stand Phytotoxicity, and Yield of Corn*

TREATMENTAND RATE/ACRE[z]	STAND COUNT #ACRE[y] MAY 24	STAND COUNT #ACRE[y] MAY 30	STAND COUNT #ACRE[y] JUN 28	PHYTO[x] %	HARVEST MOISTURE %	TEST WEIGHT LB/BU	YIELD[w] BU/ ACRE
Nontreated control	31363	35864	38478	25.0	21.6	53.3	175.6
W8S11-R003 117 SC 6.0 fl oz in-furrow	31145	37752	37026	10.0	20.6	55.0	180.3
Zironar 7.5 SC 6.0 fl oz in-furrow	32525	38115	36808	13.8	19.1	55.9	183.4
XSK03 1.98 SC 4.0 fl oz in-furrow	30056	37171	37825	8.8	20.4	54.1	180.8
Ethos XB 1.5 SC 4.0 fl oz in-furrow	30565	35647	38405	11.8	20.2	55.4	180.4
P-value[v]	0.5082	0.0929	0.5507	0.1845	0.1544	0.3396	0.4177

[z] All plots were inoculated with *Fusarium graminearum* at 1.25 g/foot within the seedbed at planting. In-furrow treatments were applied at 10 gal/acre.

[y] Stand counts were taken on May 24, May 30, and June 28 at first trifoliolate (V1), second trifoliolate (V2), and fourth trifoliolate (V4) growth stages, respectively.

[x] Plant phytotoxicity (yellowing) was visually rated as a percentage (0–100%) per plot on May 30.

[w] Yields were adjusted to 15.5% moisture and harvested on October 18.

[v] All data were analyzed in SAS 9.4 (SAS Institute, Cary, NC). A generalized linear mixed model analysis of variance was performed using PROC GLIMMIX. Values are least squares means, and values with different letters are significantly different based on the least squares difference test (α = 0.05).

EVALUATION OF FUNGICIDES FOR FOLIAR DISEASES IN CORN IN CENTRAL INDIANA, 2023 (COR23-42.ACRE)

E. A. Duncan, S. Shim, and D. E. P. Telenko, Department of Botany and Plant Pathology, Purdue University West Lafayette, IN 47907-2054

CORN (ZEA MAYS P0574AM)

Tar spot, *Phyllachora maydis*
Gray leaf spot, *Cercospora zeae maydis*
Northern corn leaf blight, *Exserohilum turcicum*

A trial was established at the Purdue Agronomy Center for Research and Education (ACRE) in Tippecanoe County, Indiana. The experiment was a randomized complete block design with four replications. Plots were 10 feet wide and 30 feet long and consisted of four rows, and the two center rows were used for evaluation. The field was no-till and had been continuously corn for over five years. Standard practices for nonirrigated grain corn production in Indiana were followed. Corn hybrid P0574AM was planted in 30-inch row spacing at a rate of 2 seeds/foot on May 11. Foliar applications were made at V10 on July 19 and silk (R1) growth stage on August 1. All foliar fungicide applications were applied at 15 gal/acre/acre and 40 psi using a Lee self-propelled sprayer equipped with a 10-foot boom, fitted with six TJ-VS 8002 nozzles spaced 20 inches apart. Disease ratings were assessed on September 14 at R5 (dent) growth stage. Tar spot, gray leaf spot (GLS), and northern corn leaf blight (NCLB) severity were visually assessed as a percentage (0–100%) of symptomatic leaf area on ear leaf of five plants per plot and averaged before analysis. Percent of canopy greenness was rated by visually assessing the percentage (0–100%) of the whole plot for crop canopy that remained green at dent (R5) growth stage on September 14. The two center rows of each plot were harvested on October 18, and yields were adjusted to 15.5% moisture. All data were analyzed in SAS 9.4 (SAS Institute, Cary, NC). A generalized linear mixed model analysis of variance was performed using PROC GLIMMIX. Values are least squares means, and values with different letters are significantly different based on the least squares difference test ($\alpha = 0.05$).

In 2023, weather conditions were unfavorable for disease development. GLS, tar spot, and NCLB were present in the trial and reached low levels. All foliar treatments significantly reduced tar spot on the ear leaf over the nontreated control except Toguard at R1 (Table 6). All treatments reduced GLS over the nontreated control except Lucento at R1 (Table 6). All fungicide programs increased canopy greenness significantly over the nontreated control except Lucento at R1 and Topguard at V10 followed by Adastrio at R1 (Table 6). No significant differences between fungicide treatments and the nontreated control were detected for NCLB, moisture, test weight, and yield of corn.

TABLE 6. *Effect of Fungicide on Foliar Disease Severity, Canopy Greenness, and Yield of Corn*

TREATMENT, RATE/ACRE, AND TIMING[z]	TAR SPOT[y] %	GLS[y] %	NCLB[y] %	CANOPY GREEN[x] %	HARVEST MOISTURE %	TEST WEIGHT LB/BU	YIELD[w] BU/ACRE
Nontreated control	0.13 a	0.8 a	2.2	86.3 c	21.2	53.4	192.0
Lucento 4.17 SC 5.0 fl oz at R1	0.05 bc	0.6 ab	1.2	87.5 bc	21.2	54.1	191.2
Adastrio 4.0 SC 8.0 fl oz at R1	0.03 bc	0.2 c	0.7	91.3 a	21.5	54.0	194.3
Topguard EQ 4.29 SC 5.0 fl oz at R1	0.08 ab	0.3 bc	1.6	90.0 ab	22.1	53.1	193.5
Topguard EQ 4.29 SC 10.0 fl oz at V10 fb Adastrio 4.0 SC 8.0 fl oz at R1	0.00 c	0.1 c	0.8	88.8 abc	21.8	53.4	193.8
Veltyma 3.34 SC 7.0 fl oz at R1	0.03 bc	0.1 c	0.2	90.0 ab	22.5	53.3	187.2
Delaro Complete 458 SC 8.0 fl oz at R1	0.03 bc	0.2 bc	0.3	91.3 a	21.8	53.9	198.2
Trivapro 2.21 SE 13.7 fl oz at R1	0.00 c	0.2 bc	1.8	91.3 a	22.4	53.1	197.0
P-value[v]	*0.0141*	*0.0175*	*0.5775*	*0.0290*	*0.2872*	*0.2822*	*0.4768*

[z] Foliar applications were made at V10 on July 19 and silk (R1) growth stage on August 1. All foliar applications were applied at 15 gal/acre, and all applications at R1 contained a nonionic surfactant (Preference) at a rate of 0.25% v/v.

[y] Foliar disease severity was visually assessed as a percentage (0–100%) of symptomatic leaf area on ear leaf, with five plants were assessed per plot and averaged before analysis on September 14 at dent (R5) growth stage. GLS = gray leaf spot, NCLB = northern corn leaf blight, fb = followed by.

[x] Canopy greenness was visually assessed as a percentage (0–100%) of crop canopy green on September 14.

[w] Yields were adjusted to 15.5% moisture and harvested on October 18.

[v] All data were analyzed in SAS 9.4 (SAS Institute, Cary, NC). A generalized linear mixed model analysis of variance was performed using PROC GLIMMIX. Values are least squares means, and values with different letters are significantly different based on the least squares difference test (α = 0.05).

COMPARISON OF FUNGICIDES FOR FOLIAR DISEASES OF SOYBEANS IN CENTRAL INDIANA, 2023 (SOY23-01.ACRE)

E. A. Duncan, S. Shim, and D. E. P. Telenko, Department of Botany and Plant Pathology, Purdue University West Lafayette, IN 47907-2054

SOYBEAN (*GLYCINE MAX* P29A19E)

Frogeye leaf spot, *Cercospora sojina*
Septoria brown spot, *Septoria glycines*
Cercospora leaf blight, *Cercospora* spp.

A trial was conducted at the Purdue Agronomy Center for Research and Education (ACRE) in Tippecanoe County, Indiana. The experiment was a randomized complete block design with four replications. Plots were 10 feet wide and 30 feet long and consisted of four rows, and the two center rows were utilized for evaluation. The previous crop was corn. Standard practices for soybean production in Indiana were followed. Soybean cultivar P29A19E was planted in 30-inch row spacing at a rate of 140,000 seeds/acre on May 11. Fungicide applications were applied on July 19 at beginning pod (R3) growth stage and were applied at 15 gal/acre at 40 psi using a Lee self-propelled sprayer equipped with a 10-foot boom, fitted with six TJ-VS 8002 nozzles spaced 20 inches apart. Disease ratings were assessed on September 11 at full seed/beginning maturity (R6/R7) growth stage. Frogeye leaf spot (FLS), Septoria brown spot (SBS), and Cercospora leaf blight (CLB) were rated by visually assessing the percentage of symptomatic leaf area. FLS and SBS were rated only in the upper and lower canopies, respectively. Percent of canopy greenness was visually assessed as a percentage (0–100%) on September 11. The two center rows of each plot were harvested on October 3, and yields were adjusted to 13% moisture. All data were analyzed in SAS 9.4 (SAS Institute, Cary, NC). A generalized linear mixed model analysis of variance was performed using PROC GLIMMIX. Values are least squares means, and values with different letters are significantly different based on the least squares difference test (α = 0.05).

In 2023, weather conditions were unfavorable for disease development. FLS, SBS, and CLB were present in the trial but only reached low levels. There was no significant difference between treatments compared to the nontreated control for FLS, SBS, and CLB severity (Table 7). Applications of Revytek resulted in significantly higher canopy greenness than the nontreated control. There was no significant difference between fungicide applications and the nontreated control for yield of soybean.

TABLE 7. *Effect of Treatment on Foliar Disease Incidence, Canopy Greenness, and Yield of Soybean*

TREATMENT AND RATE/ACRE[z]	FLS[y] %	SBS[y] %	CLB[y] %	CANOPY GREEN[x] %	YIELD[w] BU/ACRE
Nontreated control	0.3	2.5	0.3	40.0 bcd	53.5
Topguard EQ 4.29 SC 5.0 fl oz	0.6	1.0	0.0	33.8 d	50.4
Lucento 4.17 SC 5.0 fl oz	0.6	1.8	0.0	47.5 a-d	58.3
Trivapro 2.21 SE 13.7 fl oz	0.6	1.0	0.5	52.5 abc	53.5
Quadris 2.08 SC 6.0 fl oz	0.8	2.8	0.0	41.3 bcd	50.5
Veltyma 3.34 SC 7.0 fl oz	0.3	1.8	0.0	50.0 abc	54.7
Revytek 3.33 LC 8.0 fl oz	1.1	1.8	0.3	62.5 a	54.7
Echo 2.21 SE 36.0 fl oz + Folicur 3.6 F 4.0 fl oz + Topsin 4.5 FL 20.0 fl oz	0.8	2.3	0.0	55.0 ab	53.8
Delaro Complete 458 SC 8.0 fl oz	0.6	1.8	0.3	37.5 cd	50.2
Miravis Neo 2.5 SE 13.7 fl oz	1.0	3.0	0.0	47.5 a-d	53.8
Topsin 4.5 FL 20.0 fl oz	0.6	1.8	0.3	40.0 bcd	55.0
P-value[v]	0.6422	0.0846	0.7000	0.0327	0.7721

[z] Fungicide applications were made on July 19 at beginning pod/full pod (R3/R4) growth stage and contained a nonionic surfactant (Preference) at a rate of 0.25% v/v.

[y] Foliar disease severity was rated by visually assessing the percentage of symptomatic leaf area in the upper and lower canopies on September 11 at full seed/beginning maturity (R6/R7) growth stage. FLS was only rated in the upper canopy, and SBS was rated only in the lower canopy. FLS = frogeye leaf spot, SBS = Septoria brown spot, CLB = Cercospora leaf blight.

[x] Canopy greenness was visually rated on a scale of 0–100% on September 11.

[w] Yields were adjusted to 13% moisture and harvested on October 3.

[v] All data were analyzed in SAS 9.4 (SAS Institute, Cary, NC). A generalized linear mixed model analysis of variance was performed using PROC GLIMMIX. Values are least squares means, and values with different letters are significantly different based on the least squares difference test (α = 0.05).

FUNGICIDE COMPARISON IN SOYBEAN IN CENTRAL INDIANA, 2023 (SOY23-03.ACRE)

M. S. Mizuno, S. Shim, and D. E. P. Telenko, Department of Botany and Plant Pathology, Purdue University West Lafayette, IN 47907-2054

SOYBEAN (*GLYCINE MAX* P29A19E)

Frogeye leaf spot, *Cercospora sojina*
Septoria brown spot, *Septoria glycines*
Cercospora leaf blight, *Cercospora kikuchii*

A trial was established at the Purdue Agronomy Center for Research and Education (ACRE) in Tippecanoe County, Indiana. The experiment was a randomized complete block design with four replications. Plots were 10 feet wide and 30 feet long and consisted of four rows, and the two center rows were used for evaluation. The previous crop was corn. Standard practices for soybean production in Indiana were followed. Soybean cultivar P29A19E was planted in 30-inch row spacing at a rate of 140,000 seeds/acre on May 11. Fungicide applications were applied on July 21 and August 1 at beginning pod (R3) and beginning seed (R5) growth stages, respectively. All foliar fungicide applications were applied at 15 gal/acre and at 40 psi using a Lee self-propelled sprayer equipped with a 10-foot boom, fitted with six TJ-VS 8002 nozzles spaced 20 inches apart. Disease ratings were assessed on August 22 at full seed (R6) growth stage. Frogeye leaf spot (FLS), Septoria brown spot (SBS), and Cercospora leaf blight (CLB) were rated by visually assessing the percentage of symptomatic leaf area. FLS and SBS were rated only in the upper and lower canopies, respectively. The two center rows of each plot were harvested on October 3, and yields were adjusted to 13% moisture. All data were analyzed in SAS 9.4 (SAS Institute, Cary, NC). A generalized linear mixed model analysis of variance was performed using PROC GLIMMIX. Values are least squares means, and values with different letters are significantly different based on the least squares difference test (α = 0.05).

In 2023, weather conditions were not favorable for the diseases. FLS, CLB, and SBS were present in the trial but only reached low levels. There was no significant effect of treatment on FLS, CLB, and SBS severity (Table 8). There was no significant effect of treatment on canopy greenness, test weight, and yield of soybean.

TABLE 8. *Effect of Fungicide on Foliar Disease Severity, Canopy greenness, and Yield of Soybean*

TREATMENT, RATE/ACRE, AND TIMING[z]	FLS[Y] %	CLB[Y] %	SBS[Y] %	CANOPY GREEN[x] %	TEST WEIGHT LB/BU	YIELD[w] BU/ACRE
Nontreated control	0.00	0.00	0.18	58.8	57.0	58.1
Delaro Complete 458 SC 8.0 fl oz at R3	0.03	0.00	0.13	52.5	56.5	57.5
Lucento 4.17 SC 5.0 fl oz at R3	0.00	0.00	0.15	58.8	56.7	57.3
Trivapro 2.21 SE 13.7 fl oz at R3	0.00	0.00	0.45	46.3	56.5	56.9
Miravis Neo 2.5 SE 13.7 fl oz at R3	0.00	0.00	0.10	44.2	56.6	55.8
Revytek 3.33 LC 8.0 fl oz at R3	0.00	0.00	0.10	66.3	56.6	59.4
Delaro Complete 458 SC 8.0 fl oz at R5	0.00	0.00	0.13	76.3	56.7	55.4
Lucento 4.17 SC 5.0 fl oz at R5	0.01	0.00	0.15	47.5	56.3	58.7
Trivapro 2.21 SE 13.7 fl oz at R5	0.25	0.00	0.13	45.0	56.8	54.6
Miravis Neo 2.5 SE 13.7 fl oz at R5	0.03	0.00	0.10	56.3	56.4	61.4
Revytek 3.33 LC 8.0 fl oz at R5	0.01	0.01	0.18	72.5	56.5	59.5
Nontreated control	0.03	0.03	0.18	55.0	56.5	59.3
P-value[v]	*0.5284*	*0.5735*	*0.6386*	*0.7400*	*0.5744*	*0.4403*

[z] Fungicide applications were made on July 21 and August 1 at beginning pod (R3) and beginning seed (R5) growth stages, respectively. All treatments contained a nonionic surfactant (Preference) at a rate of 0.25% v/v.

[y] Foliar disease severity was rated by visually assessing as a percentage of symptomatic leaf area in the upper and lower canopies on August 22 at full seed (R6) growth stage. FLS was only rated in the upper canopy, and SBS was rated only in the lower canopy. FLS = frogeye leaf spot, SBS = Septoria brown spot, CLB = Cercospora leaf blight.

[x] Canopy greenness visually assessed on a scale of 0–100% canopy greenness within a plot on August 22.

[w] Yields were adjusted to 13% moisture and harvested on October 3.

[v] All data were analyzed in SAS 9.4 (SAS Institute, Cary, NC). A generalized linear mixed model analysis of variance was performed using PROC GLIMMIX. Values are least squares means, and values with different letters are significantly different based on the least squares difference test (α = 0.05).

COMPARISON OF PLANTING DATES AND SEED TREATMENT ON SOYBEAN IN CENTRAL INDIANA, 2023 (SOY23-10.ACRE)

I. L. Miranda, S. Shim, and D. E. P. Telenko, Department of Botany and Plant Pathology, Purdue University West Lafayette, IN 47907-2054

SOYBEAN (*GLYCINE MAX* 25E334N)

Sudden death syndrome, *Fusarium virguliforme*

A trial was established at the Purdue Agronomy Center for Research and Education (ACRE) in Tippecanoe County, Indiana. The experiment was a randomized complete block design with four replications. Plots were 10 feet wide and 30 feet long and consisted of four rows, and the two center rows were used for evaluation. The previous crop was corn. Standard practices for soybean production in Indiana were followed. Soybean seeds were planted in 30-inch row spacing at a rate of 140,000 seeds/acre. Treatments were a factorial arrangement of four planting dates by four seed treatments. Soybeans were planted on April 14 (planting date 1), April 27 (planting date 2), May 11 (planting date 3), and May 31 (planting date 4). Stand counts were assessed at cotyledons expanded/first-node stage (VC/V1) growth stage for each planting date. Disease ratings were assessed on August 22 at beginning seed (R5) growth stage. Sudden death syndrome (SDS) was rated for disease severity by visually assessing the percentage of canopy with symptoms. Ten roots were sampled from the outer rows of each plot on September 8 and rated for root rot severity on a scale of 0–100% and averaged before analysis. The two center rows of each plot were harvested on October 3, and yields were adjusted to 13% moisture. All data were analyzed in SAS 9.4 (SAS Institute, Cary, NC). A generalized linear mixed model analysis of variance was performed using PROC GLIMMIX. Values are least squares means, and values with different letters are significantly different based on a the least squares difference test (α = 0.05).

In 2023, very little disease developed in plots. SDS was present in the trial and reached low severity. No significant interactions between planting dates and seed treatments were detected; therefore, main effects of planting dates and seed treatments are presented (Table 9). Soybean stand counts were the highest at planting dates on April 14 and May 11as compared to April 27 and May 31. Planting soybean on April 14 resulted in the highest incidence of SDS compared to later planting dates. Root rot severity was significantly lower at the planting on May 31 compared to all the other planting dates. No differences were detected between planting dates and seed treatments on root weight. Test weight at planting on May 11 resulted in the highest compared to all other planting dates. Soybean yield was significantly reduced higher in plantings on April 14 and April 27 compared to plantings on May 11 and May 31. No significant differences were detected between seed treatments for stand count, SDS% incidence, root rot, root dry weight, test weight, and soybean yield.

TABLE 9. *Effect of Planting Dates and Seed Treatments on Stand Count, SDS, Root Rot, Root Weight and Yield of Soybean*

PLANTING DATES AND SEED TREATMENTS[z]	STAND COUNT #/ACRE	SDS DI[y] %	ROOT ROT[x] %	ROOT DRY WEIGHT[w] G	TEST WEIGHT LB/BU	YIELD[v] BU/ACRE
Planting date 1 (April 14)	98,337 a	5.0 a	0.3 a	27.0	56.0 b	69.2 a
Planting date 2 (April 27)	79,933 b	1.2 b	0.2 a	27.6	56.0 b	71.6 a
Planting date 3 (May 11)	99,698 a	0.3 b	0.2 a	27.4	56.6 a	63.3 b
Planting date 4 (May 31)	75,958 b	0.0 b	0.1 b	25.6	56.1 b	56.6 c
Nontreated control	88,917	1.1	0.2	26.6	56.2	65.5
CruiserMaxx APX (with Thiamethoxam)	91,639	2.2	0.2	27.7	56.1	65.1
Thiamethoxam	80,041	1.2	0.2	28.4	56.0	66.9
CruiserMaxx APX (without Thiamethoxam)	93,327	2.1	0.2	24.9	56.4	63.1
P-value *planting date*[u]	*0.0001*	*0.0049*	*0.0163*	*0.6030*	*0.0063*	*0.0001*
P-value *seed treatment*	*0.0511*	*0.8141*	*0.5660*	*0.1766*	*0.1714*	*0.4386*
P-value *planting date*seed treatment*	*0.4390*	*0.9701*	*0.9610*	*0.1728*	*0.7819*	*0.4004*

[z] Seed treatments applied prior to planting at 10 g AI/100 kg seed.

[y] Disease incidence was visually assessed as a percentage (0–100%) of canopy with disease symptoms on August 22. SDS = sudden death syndrome.

[x] Root rot was visually assessed as a percentage (0–100%) of dark discoloration on roots on September 8.

[w] Root dry weight = weight of 10 dried root samples in grams (g) on September 21.

[v] Yields were adjusted to 13% moisture and harvested on October 3.

[u] All data were analyzed in SAS 9.4 (SAS Institute, Cary, NC). A generalized linear mixed model analysis of variance was performed using PROC GLIMMIX. Values are least squares means, and values with different letters are significantly different based on the least squares difference test (α = 0.05).

EVALUATION OF FUNGICIDES FOR SOYBEAN FOLIAR DISEASES IN CENTRAL INDIANA, 2023 (SOY23-16.ACRE)

E. A. Duncan, S. Shim, and D. E. P. Telenko, Department of Botany and Plant Pathology, Purdue University West Lafayette, IN 47907-2054

SOYBEAN (*GLYCINE MAX* P29A19E)

Frogeye leaf spot, *Cercospora sojina*
Cercospora leaf blight, *Cercospora kikuchii*
Septoria brown spot, *Septoria glycines*

A trial was established at the Agronomy Center for Research and Education (ACRE) in Tippecanoe County, Indiana. The trial was a randomized complete block design with four replications. Plots were 10 feet wide and 30 feet long and consisted of four rows, and the two center rows were used for evaluation. The previous crop was corn. Standard practices for soybean production in Indiana were followed. Soybean cultivar P29A19E was planted in 30-inch row spacing at a rate of 140,000 seeds/acre on May 11. Fungicides were applied on July 21 at beginning pod (R3) growth stage. All fungicide applications were applied at 15 gal/acre and at 40 psi using a Lee self-propelled sprayer equipped with a 10-foot boom, fitted with six TJ-VS 8002 nozzles spaced 20 inches apart. Disease ratings were assessed on September 11 at full seed/beginning maturity (R6/R7) growth stage. Frogeye leaf spot (FLS), Septoria brown spot (SBS), and Cercospora leaf blight (CLB) were rated by visually assessing the percentage of symptomatic leaf area. FLS and SBS were rated only in the upper and lower canopies, respectively. Percent of canopy green was visually assessed as a percentage (0–100%) of crop canopy green on September 14. The two center rows of each plot were harvested on October 3, and yields were adjusted to 13% moisture. All data were analyzed in SAS 9.4 (SAS Institute, Cary, NC). A generalized linear mixed model analysis of variance was performed using PROC GLIMMIX. Values are least squares means, and values with different letters are significantly different based on the least squares difference test ($\alpha = 0.05$).

In 2023, weather conditions were not favorable for diseases. FLS, SBS, and CLB were present in the trial but only reached low levels. There was no significant effect on treatments on FLS and CLB severities compared to the nontreated control (Table 10). All fungicide applications significantly reduced SBS severity in the lower canopy when compared to the nontreated control. Quadris Top resulted in the lowest severity of SBS, but it was not significantly different from other fungicide applications except for ADM.03509.F.1.A at 4.8 fl oz, ADM.03509.F.1.A at 8.0 fl oz, and Stratego. There was no significant difference between all fungicide applications and the nontreated control for canopy greenness, test weight, and yield of soybean.

TABLE 10. *Effect of Treatment on Foliar Disease Severity, Canopy Greenness, and Yield of Soybean*

TREATMENT AND RATE/ACRE[z]	FLS[Y] %	SBS[Y] %	CLB[Y] %	CANOPY GREEN[x] %	TEST WEIGHT LB/BU	YIELD[w] BU/ACRE
Nontreated control	0.3	5.5 a	0.3	52.5	56.5	53.5
ADM.03509.F.3.B 1.75 EC 8.0 fl oz	0.1	2.0 cd	0.0	57.5	56.1	54.5
ADM.03509.F.3.B 1.75 EC 16.0 fl oz	0.2	1.8 cd	0.1	55.0	56.9	53.3
ADM.03509.F.1.A 2.92 SC 4.8 fl oz	0.2	2.8 bc	0.1	52.5	56.5	52.7
ADM.03509.F.1.A 2.92 SC 9.6 fl oz	0.1	2.3 cd	0.0	57.5	56.3	56.4
ADM.03509.F.1.A 2.92 SC 8.0 fl oz	0.1	3.0 bc	0.3	57.5	56.6	53.1
ADM.03509.F.1.A 2.92 SC 8.0 fl oz + Liberate 0.25% v/v	0.1	2.5 bcd	0.0	58.8	56.2	56.7
Custodia 2.67 SC 6.8 fl oz	0.3	2.6 bcd	0.3	61.3	56.2	56.5
Mattock 50 WP 6.8 fl oz	0.2	2.6 cd	0.0	55.0	56.6	55.1
Stratego 2.08 SC 4.0 fl oz	0.1	3.6 b	0.0	58.8	56.2	56.7
Quadris Top 1.67 SC 9.0 fl oz	0.1	1.4 d	0.2	50.0	56.7	53.2
P-value[v]	*0.3367*	*0.0001*	*0.7156*	*0.7101*	*0.4360*	*0.8556*

[z] Fungicides were applied on July 21 at beginning pod (R3) growth stage.

[y] Foliar disease severity was rated by visually assessing the percentage of symptomatic leaf area in the upper and lower canopies at full seed/beginning maturity (R6/R7) growth stage on September 11. FLS was only rated in the upper canopy, and SBS was rated only in the lower canopy. FLS = frogeye leaf spot, SBS = Septoria brown spot, CLB = Cercospora leaf blight.

[x] Canopy greenness was visually assessed as a percentage (0–100%) of crop canopy green on September 11.

[w] Yields were adjusted to 13% moisture and harvested on October 3.

[v] All data were analyzed in SAS 9.4 (SAS Institute, Cary, NC). A generalized linear mixed model analysis of variance was performed using PROC GLIMMIX. Values are least squares means, and values with different letters are significantly different based on the least squares difference test (α = 0.05).

EVALUATION OF IN-FURROW FUNGICIDE EFFICACY FOR SOYBEAN CYST NEMATODES IN CENTRAL INDIANA, 2023 (SOY23-18.ACRE)

E. A. Duncan, S. Shim, and D. E. P. Telenko, Department of Botany and Plant Pathology, Purdue University West Lafayette, IN 47907-2054

SOYBEAN (*GLYCINE MAX* P29A19E)

Sudden death syndrome, *Fusarium virguliforme*
Soybean cyst nematode, *Heterodera glycines*

A trial was established at the Agronomy Center for Research and Education (ACRE) in Tippecanoe County, Indiana. The trial was a randomized complete block design with four replications. Plots were 10 feet wide and 30 feet long and consisted of four rows, and the two center rows were used for evaluation. The previous crop was corn. Standard practices for soybean production in Indiana were followed. Soybean cultivar P29A19E was planted in 30-inch row spacing at a rate of 140,000 seeds/acre on May 31. In-furrow applications were applied at planting in 10 gal/acre. Disease ratings were assessed on September 11 at full seed/beginning maturity (R6/R7) growth stage. Sudden death syndrome (SDS) in each plot was rated for disease incidence (DI) as percentage of plants with disease symptoms (0–100%) and disease severity (DS) on a scale of 1 to 9, where 1 refers to low disease pressure and 9 refers to premature death of the plant. The SDS index (DX) was calculated using the equation DX = (DI*DS)/9. Soil samples were taken on June 8 at nine weeks after planting (9 WAP) and September 29 at harvest for soybean cyst nematode (SCN) egg count and processed by SCN Diagnostics, Columbia, MO. The two center rows of each plot were harvested on October 4, and yields were adjusted to 13% moisture. All data were analyzed in SAS 9.4 (SAS Institute, Cary, NC). All data were analyzed in SAS 9.4 (SAS Institute, Cary, NC). A generalized linear mixed model analysis of variance was performed using PROC GLIMMIX. Values are least squares means, and values with different letters are significantly different based on the least squares difference test (α = 0.05).

In 2023, weather conditions were unfavorable for disease development. SDS and SCN were present in the trial but reached low levels. There was no significant effect of treatment on SDS (Table 11). There was no significant effect of treatment on SCN egg counts, test weight, and yield of soybean.

TABLE 11. *Effect of Treatment on SDS, Soybean Cyst Nematode (SCN) eggs, and Yield of Soybean*

TREATMENT AND RATE/ACRE[z]	SDS DI[y]	SDS INDEX[x]	SCN EGGS[w] 9 WAP	SCN EGGS[w] HARVEST	TEST WEIGHT LB/BU	YIELD[v] BU/ACRE
Nontreated control	1.0	0.4	213	350	54.6	64.3
AMV1310 0.3 EC 14.1 fl oz	0.8	0.2	100	600	55.4	63.4
AMV1584 4.16 SC 6.84 fl oz	0.3	0.1	825	725	55.2	61.2
AMV1984 0.08 L 16.0 fl oz	0.3	0.1	188	475	55.7	64.8
AMV2185 0.037 D 6.0 oz	0.8	0.2	650	500	55.2	61.8
AMV1306 2.55 L 20.0 fl oz	0.8	0.4	175	875	55.6	63.5
P-value[u]	0.6645	0.3921	0.1387	0.8408	0.2533	0.5383

[z] In-furrow treatments applied at planting at 10 gal/acre.

[y] SDS in each plot was rated for disease incidence (DI) as a percentage of plants with disease symptoms (0–100%) on September 11.

[x] SDS Index (DX) calculated using the equation DX= (DI*DS/9). SDS = sudden death syndrome.

[w] SCN counts determined in 100 cc soil and processed by SCN Diagnostics, Columbia, MO, on July 18 at 9 WAP (weeks after planting) and September 29 at harvest.

[v] Yields were adjusted to 13% moisture and harvested on October 4.

[u] All data were analyzed in SAS 9.4 (SAS Institute, Cary, NC). A generalized linear mixed model analysis of variance was performed using PROC GLIMMIX. Values are least squares means, and values with different letters are significantly different based on the least squares difference test (α = 0.05).

EVALUATION OF FUNGICIDES FOR SOYBEAN FOLIAR DISEASES IN CENTRAL INDIANA, 2023 (SOY23-23.ACRE)

E. A. Duncan, S. Shim, and D. E. P. Telenko, Department of Botany and Plant Pathology, Purdue University West Lafayette, IN 47907-2054

SOYBEAN (*GLYCINE MAX* P29A19E)

Frogeye leaf spot, *Cercospora sojina*
Cercospora leaf blight, *Cercospora kikuchii*
Septoria brown spot, *Septoria glycines*

A trial was established at the Purdue Agronomy Center for Research and Education (ACRE) in Tippecanoe County, Indiana. The trial was a randomized complete block design with four replications. Plots were 10 feet wide and 30 feet long and consisted of four rows, and the two center rows were used for evaluation. The previous crop was corn. Standard practices for soybean production in Indiana were followed. Soybean cultivar P29A19E was planted in 30-inch row spacing at a rate of 140,000 seeds/acre on May 11. Fungicide applications were applied on June 28 at fourth trifoliate (V4) and on July 19 at beginning pod (R3) growth stages. All foliar fungicide applications were applied at 15 gal/acre and 40 psi using a Lee self-propelled sprayer equipped with a 10-foot boom, fitted with six TJ-VS 8002 nozzles spaced 20 inches apart. Disease ratings were assessed on September 11 at full seed/beginning maturity (R6/R7) growth stage. Frogeye leaf spot (FLS), Septoria brown spot (SBS), and Cercospora leaf blight (CLB) were rated by visually assessing the percentage of symptomatic leaf area. FLS and SBS were rated only in the upper and lower canopies, respectively. The two center rows of each plot were harvested on October 3, and yields were adjusted to 13% moisture. All data were analyzed in SAS 9.4 (SAS Institute, Cary, NC). A generalized linear mixed model analysis of variance was performed using PROC GLIMMIX. Values are least squares means, and values with different letters are significantly different based on the least squares difference test (α = 0.05).

In 2023, weather conditions were unfavorable for disease development. FLS, SBS, CLB were present in the trial but only reached low levels. There was no significant effect of foliar fungicide treatments on FLS, CLB, and SBS severities (Table 12). There was no significant effect of treatment compared to the nontreated control on harvest moisture, test weight, and yield of soybean.

TABLE 12. *Effect of Treatment on Foliar Diseases and Yield of Soybean*

TREATMENT AND RATE/ACRE[z]	FLS[y] %	SBS[y] %	CLB[y] %	HARVEST MOISTURE %	TEST WEIGHT LB/BU	YIELD[x] BU/ACRE
Nontreated control	0.7	3.1	0.0	9.8	56.9	58.2
Topguard 1.04 SC 7.0 fl oz at V4 fb Lucento 4.17 SC 5.0 fl oz at R3	0.2	2.0	0.2	9.9	56.8	62.2
Adastrio 4.0 SC 7.0 fl oz at R3	0.2	1.5	0.0	9.9	57.0	58.7
Miravis Top 1.67 SC 13.7 fl oz at R3	0.2	1.7	0.0	9.9	56.8	58.5
Revytek 3.33 LC 8.0 fl oz at R3	0.1	0.9	0.0	10.0	57.0	57.3
Lucento 4.17 SC 5.0 fl oz at R3	0.1	1.2	0.0	9.8	56.9	57.7
P-value[w]	*0.3289*	*0.4726*	*0.3511*	*0.1543*	*0.9619*	*0.5271*

[z] Fungicide applications were made on June 28 at fourth trifoliate (V4) and on July 21 at beginning pod (R3) growth stages. All applications contained a nonionic surfactant (Preference) at a rate of 0.25% v/v. fb = followed by.

[y] Foliar disease severity was rated by visually assessing the percentage of symptomatic leaf area in the upper and lower canopies on September 11 at full seed/beginning (R6/R7) growth stage. FLS was only rated in the upper canopy, and SBS was rated only in the lower canopy. FLS = frogeye leaf spot, SBS = Septoria brown spot, CLB = Cercospora leaf blight.

[x] Yields were adjusted to 13% moisture and harvested on October 4.

[w] All data were analyzed in SAS 9.4 (SAS Institute, Cary, NC). A generalized linear mixed model analysis of variance was performed using PROC GLIMMIX. Values are least squares means, and values with different letters are significantly different based on the least squares difference test (α = 0.05).

FUNGICIDE COMPARISON IN SOYBEAN IN CENTRAL INDIANA, 2023 (SOY23-28.ACRE)

E. A. Duncan, S. Shim, and D. E. P. Telenko, Department of Botany and Plant Pathology, Purdue University West Lafayette, IN 47907-2054

SOYBEAN (*GLYCINE MAX* P29A19E)

Frogeye leaf spot, *Cercospora sojina*
Cercospora leaf blight, *Cercospora kikuchii*
Septoria brown spot, *Septoria glycines*

A trial was established at the Purdue Agronomy Center for Research and Education (ACRE) in Tippecanoe County, Indiana. The trial was a randomized complete block design with four replications. Plots were 10 feet wide and 30 feet long and consisted of four rows, and the two center rows were used for evaluation. The previous crop was corn. Standard practices for soybean production in Indiana were followed. Soybean cultivar P29A19E was planted in 30-inch row spacing at a rate of 140,000 seeds/acre on May 11. Fungicide applications were applied on July 21 at beginning pod (R3) growth stage. All foliar fungicides were applied at 15 gal/acre at 40 psi using a Lee self-propelled sprayer equipped with a 10-foot boom, fitted with six TJ-VS 8002 nozzles spaced 20 inches apart. Disease ratings were assessed on September 11 at full seed/beginning maturity (R6/R7) growth stage. Frogeye leaf spot (FLS), Septoria brown spot (SBS), and Cercospora leaf blight (CLB) were rated for disease severity by visually assessing the percentage of symptomatic leaf area. FLS and SBS were rated only in the upper and lower canopies, respectively. The two center rows of each plot were harvested on September 25, and yields were adjusted to 13% moisture. All data were analyzed in SAS 9.4 (SAS Institute, Cary, NC). A generalized linear mixed model analysis of variance was performed using PROC GLIMMIX. Values are least squares means, and values with different letters are significantly different based on the least squares difference test (α = 0.05).

In 2023, weather conditions were unfavorable for disease development. FLS, SBS, and CLB were present in the trial but only reached low levels. There was no significant effect of treatment on FLS and CLB severity (Table 13). All fungicide applications significantly reduced SBS over the nontreated control on September 11. There was no significant effect of treatment on harvest moisture and test weight. Miravis Neo + Endigo and Miravis Top + Endigo significantly increased yield over the nontreated control but were not significantly different from Miravis Top and Revytek.

TABLE 13. *Effect of Treatment on Foliar Disease Severity and Yield of Soybean*

TREATMENT AND RATE/ACRE[z]	FLS[y] %	SBS[y] %	CLB[y] %Y	HARVEST MOISTURE %	TEST WEIGHT LB/BU	YIELD[x] BU/ACRE
Nontreated control	1.3	5.5 a	0.1	11.1	56.6	59.7 b
Miravis Neo 2.5 SE 13.7 fl oz	0.1	1.0 b	0.1	11.5	56.4	59.4 b
Miravis Top 1.67 SC 13.7 fl oz	0.1	0.8 b	0.0	11.0	56.5	62.2 ab
Miravis Neo 2.5 SE 13.7 fl oz + Endigo ZC 2.06 SC 4.0 fl oz	0.2	0.8 b	0.1	11.7	56.5	63.9 a
Miravis Top 1.67 SC 13.7 fl oz + Endigo ZC 2.06 SC 4.0 fl oz	0.1	1.4 b	0.0	11.2	56.7	64.8 a
Miravis Neo 2.5 SE 13.7 fl oz + Warrior II 2.08 CS 1.6 fl oz	0.1	0.9 b	0.0	10.9	56.8	58.6 b
Revytek 3.33 LC 8.0 fl oz	0.1	1.5 b	0.2	11.4	56.2	61.0 ab
P-value[w]	0.4788	0.0010	0.5405	0.3435	0.8133	0.0422

[z] Fungicide applications were made on July 21 at beginning pod (R3) growth stage and contained a nonionic surfactant (Preference) at a rate of 0.25% v/v.

[y] Foliar disease severity was rated by visually assessing the percentage of symptomatic leaf area in the upper and lower canopies on September 11 at full seed/beginning maturity (R6/R7) growth stage. FLS was only rated in the upper canopy, and SBS was rated only in the lower canopy. FLS = frogeye leaf spot, SBS = Septoria brown spot, CLB =Cercospora leaf blight.

[x] Yields were adjusted to 13% moisture and harvested on September 25.

[w] All data were analyzed in SAS 9.4 (SAS Institute, Cary, NC). A generalized linear mixed model analysis of variance was performed using PROC GLIMMIX. Values are least squares means, and values with different letters are significantly different based on the least squares difference test (α = 0.05).

FROGEYE LEAF SPOT MODEL EVALUATION FOR FUNGICIDE APPLICATION IN SOYBEAN IN CENTRAL INDIANA, 2023 (SOY23-34.ACRE)

E. A. Duncan, S. Shim, and D. E. P. Telenko, Department of Botany and Plant Pathology, Purdue University West Lafayette, IN 47907-2054

SOYBEAN (*GLYCINE MAX* P29A19E)

Frogeye leaf spot, *Cercospora sojina*
Septoria brown spot, *Septoria glycines*

A trial was established at the Purdue Agronomy Center for Research and Education (ACRE) in Tippecanoe County, IN. The trial was a randomized complete block design with four replications. Plots were 10 feet wide and 30 feet long and consisted of four rows, and the two center rows were used for evaluation. The previous crop was corn. Standard practices for soybean production in Indiana were followed. Soybean cultivar P29A19E was planted in 30-inch row spacing at a rate of 140,000 seeds/acre on May 11. Fungicide applications were applied on July 19 at beginning pod (R3) growth stage. Model applications were not made, as the model remained below all thresholds. All applications were applied at 15 gal/acre and at 40 psi using a Lee self-propelled sprayer equipped with a 10-foot boom, fitted with six TJ-VS 8002 nozzles spaced 20-ichesn apart. Disease ratings were assessed on September 11 at full seed/beginning maturity (R6/R7) growth stage. Frogeye leaf spot (FLS) and Septoria brown spot (SBS) were rated for disease severity by visually assessing the percentage (0–100%) of symptomatic leaf area. FLS and SBS were rated only in the upper and lower canopies, respectively. The two center rows of each plot were harvested on October 4, and yields were adjusted to 13% moisture. All data were analyzed in SAS 9.4 (SAS Institute, Cary, NC). A generalized linear mixed model analysis of variance was performed using PROC GLIMMIX. Values are least squares means, and values with different letters are significantly different based on the least squares difference test (α = 0.05).

In 2023, weather conditions were unfavorable for disease development. FLS and SBS were present in the trial but only reached low levels. There was no significant effect of treatment on FLS severity compared to the nontreated control (Table 14). Revytek applied at R3 reduced SBS severity, increased canopy greenness, and increased harvest moisture over the nontreated control. There was no significant effect of treatment on test weight and yield of soybean.

TABLE 14. *Effect of Treatment on Foliar Diseases and Yield of Soybean*

TREATMENT, RATE/ACRE, AND TIMING[z]	FLS[Y] %	SBS[Y] %	CANOPY GREEN[x] %	HARVEST MOISTURE %	TEST WEIGHT LB/BU	YIELD[W] BU/ACRE
Nontreated control	0.08	5.8 a	41.3 b	9.3 b	55.8	45.7
Revytek 3.33 LC 8.0 fl oz at R3	0.05	0.2 c	56.3 a	9.5 a	55.8	47.1
Revytek 3.33 LC 8.0 fl oz at threshold 50% no application	0.00	5.0 ab	43.8 b	9.3 b	56.0	45.4
Revytek 3.33 LC 8.0 fl oz at threshold 60% no application	0.00	3.0 b	46.3 b	9.4 ab	55.7	44.5
Revytek 3.33 LC 8.0 fl oz at threshold 70% no application	0.03	4.8 ab	47.5 b	9.4 ab	55.7	46.1
P-value[v]	*0.1283*	*0.0028*	*0.0160*	*0.0291*	*0.8074*	*0.7411*

[z] Fungicide applications were made on July 19 at beginning pod (R3) growth stage. All treatments contained a nonionic surfactant (Preference) at a rate of 0.25% v/v. Model never reached any of the three thresholds (50%, 60%, 70%), so only the R3 treatment was applied.

[y] Foliar disease severity was rated by visually assessing the percentage (0–100%) of symptomatic leaf area in the upper and lower canopies, respectively, on September 11 at full seed/beginning maturity (R6/R7) growth stage. FLS was only rated in the upper canopy, and SBS was rated only in the lower canopy. FLS = frogeye leaf spot, SBS = Septoria brown spot.

[x] Canopy greenness was visually assessed on a scale of 0–100% canopy greenness within a plot on September 11.

[w] Yields were adjusted to 13% moisture and harvested on October 4.

[v] All data were analyzed in SAS 9.4 (SAS Institute, Cary, NC). A generalized linear mixed model analysis of variance was performed using PROC GLIMMIX. Values are least squares means, and values with different letters are significantly different based on the least squares difference test (α = 0.05).

EVALUATION OF PRODUCTS AND CULTIVARS FOR SCAB MANAGEMENT IN ORGANIC WHEAT IN INDIANA, 2023 (WHT23-01.ACRE)

C. Rocco da Silva, S. Shim, and D. E. P. Telenko, Department of Botany and Plant Pathology, Purdue University West Lafayette, IN 47907-2054

WHEAT (*TRITICUM AESTIVUM* HARPOON AND KASKASKIA)

Fusarium head blight (Scab), *Fusarium graminearum*

A trial was established at the Purdue Agronomy Center for Research and Education (ACRE) in Tippecanoe County, Indiana. The experiment was a randomized complete block design with four replications. Plots were 7.5 feet wide and 20 feet long and consisted of 12 rows spaced 7.5 inches apart, and the center of each plot was used for evaluation. The previous crop was corn. Organic wheat cultivars Harpoon and Kaskaskia were planted in 7.5-inch row spacing using a drill on October 18, 2022. All fungicide applications were applied at 15 gal/acre and 40 psi using a CO_2 backpack sprayer equipped with a 10-foot boom, fitted with six TJ-VS 8002 nozzles spaced 20 inches apart and directed forward and backward at a 45-degree angle. Fungicides were applied on May 22 at Feekes growth stage 10.5.1. All plots were inoculated with a mixture of isolates of *Fusarium graminearum* endemic to Indiana on May 23 with a spore suspension (50,000 spores/ml) applied at 300 ml/plot with the CO_2 handheld sprayer. Disease ratings were assessed on June 10. Fusarium head blight (FHB) incidence was measured as the number of infected heads out of 60 plants in each plot and calculated as a percentage. FHB severity was rated by visually assessing the percentage (0–100%) of the infected heads. The FHB index was calculated as (% FHB incidence multiplied by % FHB severity)/100 per plot. The eight center rows of each plot were harvested with a Kincaid small-plot combine on July 10, and yields were adjusted to 13.5% moisture for comparison. A subsample of grain was taken from each plot and partitioned for DON (deoxynivalenol) analysis completed by the University of Minnesota DON testing lab and to determine Fusarium damaged kernels (FDK) by visually assessing the percentage (0–100%) of the infected heads. All data were analyzed in SAS 9.4 (SAS Institute, Cary, NC). A generalized linear mixed model analysis of variance was performed using PROC GLIMMIX. Values are least squares means, and values with different letters are significantly different based on the least squares difference test ($\alpha = 0.05$).

In 2023, weather conditions were not favorable for FHB and leaf blotch diseases. Low levels of FHB were detected in the trial. There were no significant interactions between cultivar and fungicide treatments; therefore, main effects of each are presented (Table 15). In the cultivar Harpoon, FHB incidence and severity were reduced significantly when compared to Kaskaskia. No significant difference was detected for FHB index between the cultivars. There were no significant differences in treatments from the nontreated control for FHB incidence, FHB severity, and FHB index and % FDK. The concentration of DON was not detected in grain subsamples from the trial. The cultivar Kaskaskia had higher grain yield than Kaskaskia, and no significant differences were detected between the treatments and the nontreated control.

TABLE 15. *Effect of Cultivar and Fungicide on Fusarium Head Blight (FHB), Deoxynivalenol (DON), Fusarium Damaged Kernels (FDK), and Yield of Wheat*

TREATMENT[z]	FHB DI[y] %	FHB DS[x] %	FHB INDEX[w]	DON[v] PPM	FDK[u] %	YIELD[t] BU/ACRE
Cultivar						
Harpoon	0.7 b	0.5 b	0.0	nd	0.0	82.4 b
Kaskaskia	4.3 a	6.8 a	0.3	nd	0.0	89.1 a
Fungicide rate/acre						
Nontreated control	2.3	1.9	0.1	nd	0.0	85.3
Prosaro 421 SC 8.2 fl oz	2.5	1.5	0.1	nd	0.0	88.1
ChampION 50 WP 1.5 lb	2.1	0.8	0.0	nd	0.0	84.7
Pacesetter WS 13.0 fl oz	2.3	5.4	0.3	nd	0.0	85.5
Sonata SC 1.0 qt	2.3	0.8	0.0	nd	0.0	84.8
Actinovate AG 12.0 fl oz	3.3	11.5	0.5	nd	0.0	86.2
P-value *cultivar*[s]	0.0001	0.0130	0.0509	—	—	0.0001
P-value *fungicide*	0.8036	0.0960	0.0041	—	—	0.7860
P-value *cultivar*fungicide*	0.2238	0.1238	0.0562	—	—	0.9111

[z] Fungicides were applied on May 22 at Feekes growth stage 10.5.1. All plots were inoculated with a mixture of isolates of *Fusarium graminearum* endemic to Indiana on May 24, with a spore suspension (50,000 spores/ml) applied at 300 ml/plot with CO_2 handheld sprayer on May 23.

[y] FHB DI disease incidence was measured as the number of infected heads out of 60 plants in each plot and calculated as a percentage on June 10.

[x] FHB SD disease severity was rated by visually assessing the percentage of the infected head.

[w] FHB index was calculated as (% FHB DI multiplied by % FHB DS)/100 per plot.

[v] Analysis of the mycotoxin deoxynivalenol (DON) was completed by the University of Minnesota DON Testing Lab on August 17. nd = not detected, DON <0.05 ppm.

[u] Fusarium damaged kernels (FDK) was visually assessed as a percentage (0–100%) of the infected heads

[t] Yields were adjusted to 13.5% moisture and harvested on July 10.

[s] All data were analyzed in SAS 9.4 (SAS Institute, Cary, NC). A generalized linear mixed model analysis of variance was performed using PROC GLIMMIX. Values are least squares means, and values with different letters are significantly different based on the least squares difference test (α = 0.05).

EVALUATION OF FOLIAR FUNGICIDES FOR SCAB MANAGEMENT IN CENTRAL INDIANA (WHT23-02.ACRE)

C. Rocco da Silva, S. Shim, and D. E. P. Telenko, Department of Botany and Plant Pathology, Purdue University West Lafayette, IN 47907-2054

WHEAT (*TRITICUM AESTIVUM* P25R40)

Fusarium head blight (Scab), *Fusarium graminearum*

A trial was established at the Purdue Agronomy Center for Research and Education (ACRE) in Tippecanoe County, Indiana. The trial was a randomized complete block design with four replications. Plots were 7.5 feet wide and 20 feet long and consisted of 12 rows spaced 7.5 inches apart, and the center of each plot was used for evaluation. The previous crop was corn. On November 18, 2022, wheat cultivar P25R40 was drilled at 7.5 inches spacing. All fungicide applications were applied at 15 gal/acre and at 40 psi using a CO_2 backpack sprayer equipped with a 10-foot boom, fitted with six TJ-VS 8002 nozzles spaced 20 inches apart and directed forward and backward at a 45-degree angle. Fungicides were applied on May 22 and May 27 at the Feekes growth stages 10.5.1 and 10.5.1 + 5 days, respectively. All plots were inoculated with a mixture of isolates of *Fusarium graminearum* endemic to Indiana on May 23. The spore suspension (50,000 spores/ml) was applied at 300 ml/plot with the CO_2 backpack sprayer. Disease ratings were assessed on June 10. Fusarium head blight (FHB) incidence was measured as the number of infected heads out of 60 plants in each plot and calculated as a percentage. FHB severity was rated by visually assessing the percentage of the infected head, FHB index was calculated as (% FHB incidence multiplied by % FHB severity)/100 per plot. The eight center rows of each plot were harvested with a Kincaid plot combine on July 10, and yields were adjusted to 13.5% moisture. A subsample of grain was taken from each plot and partitioned for DON (deoxynivalenol) analysis completed by the University of Minnesota DON testing lab and to determine Fusarium damaged kernels (FDK) by visually assessing the percentage (0–100%) of the infected heads. All data were analyzed in SAS 9.4 (SAS Institute, Cary, NC). A generalized linear mixed model analysis of variance was performed using PROC GLIMMIX. Values are least squares means, and values with different letters are significantly different based on the least squares difference test (α = 0.05).

In 2023, weather conditions were not favorable for FHB. No FHB, DON, or FDK were detected in the trial. Harvest moisture was highest in plots with applications of Miravis Ace applied at 10.5.1 alone or followed by either Prosaro Pro, Sphaerex, or tebuconazole as compared to the nontreated control (Table 16). Prosaro Pro had significantly higher test weight than Sphaerex or Miravis Ace followed by tebuconazole, but no treatments were significantly different from nontreated control. No differences were detected between treatments for grain yield. DON was not detected in trial.

TABLE 16. *Effect of Fungicide on Leaf Blotch, Fusarium Head Blight (FHB), Deoxynivalenol (DON), Fusarium Damaged Kernels (FDK), and Yield of Wheat*

TREATMENT AND RATE/ACRE[z]	FDK[x] %	HARVEST MOISTURE %	TEST WEIGHT LB/BU	YIELD[w] BU/ACRE
Nontreated control	0.0	14.6 d	55.7 abc	83.5
Prosaro 421 SC 6.5 fl oz at 10.5.1	0.0	14.7 cd	55.7 abc	83.9
Caramba 90 EC 13.5 fl oz at 10.5.1	0.0	14.6 cd	55.5 abc	84.7
Miravis Ace 5.2 SC 13.7 fl oz at 10.5.1	0.0	15.5 ab	55.9 ab	82.1
Prosaro Pro 400 SC 10.3 fl oz at 10.5.1	0.0	14.9 bcd	56.0 a	82.8
Sphaerex 2.50 SC 7.3 fl oz at 10.5.1	0.0	14.9 bcd	55.2 c	82.4
Miravis Ace 5.2 SC 13.7 fl oz at 10.5.1 fb				
Prosaro Pro 400 SC 10.3 fl oz at 10.5.1 + 5 d	0.0	15.7 a	55.4 bc	84.2
Miravis Ace 5.2 SC 13.7 fl oz at 10.5.1 fb				
Sphaerex 2.50 SC 7.3 fl oz at 10.5.1 + 5 d	0.0	15.6 a	55.7 abc	79.5
Miravis Ace 5.2 SC 13.7 fl oz at 10.5.1 fb				
Tebuconazole 4.0 fl oz at 10.5.1 + 5 d	0.0	15.3 abc	55.2 c	77.0
P-value[v]	—	0.0071	0.0430	0.4076

[z] Fungicide treatments applied on May 22 and May 27 at Feekes growth stage 10.5.1 and 10.5.1 f + 5 days, respectively. All plots were inoculated with a mixture of isolates of *Fusarium graminearum* endemic to Indiana on May 23 with a spore suspension (50,000 spores/ml) applied at 300 ml/plot with a CO_2 backpack sprayer on May 23. fb = followed by.

[y] Analysis of the mycotoxin deoxynivalenol (DON) was completed by the University of Minnesota DON Testing Lab on August 17.

[x] Fusarium damaged kernels (FDK) was visually assessed as a percentage (0–100%) of the infected heads.

[w] Yields were adjusted to 13.5% moisture and harvested on July 10.

[v] All data were analyzed in SAS 9.4 (SAS Institute, Cary, NC). A generalized linear mixed model analysis of variance was performed using PROC GLIMMIX. Values are least squares means, and values with different letters are significantly different based on the least squares difference test (α = 0.05).

EVALUATION OF FOLIAR FUNGICIDES AND CULTIVARS FOR SCAB MANAGEMENT IN CENTRAL INDIANA, 2023 (WHT23-03.ACRE)

C. Rocco da Silva, S. Shim, and D. E. P. Telenko, Department of Botany and Plant Pathology, Purdue University West Lafayette, IN 47907-2054

WHEAT (*TRITICUM AESTIVUM* P25R40 AND P25R61)

Fusarium head blight, *Fusarium graminearum*

A trial was established at the Purdue Agronomy Center for Research and Education (ACRE) in Tippecanoe County, Indiana. The trial was a randomized complete block design with four replications. Plots were 7.5 feet wide and 20 feet long and consisted of 12 rows spaced 7.5 inches apart, and the center of each plot was used for evaluation. The previous crop was corn. On November 18, 2022, wheat cultivar P25R40 and P25R61 were drilled at 7.5 inches. All fungicide applications were applied at 15 gal/acre and at 40 psi using a CO_2 backpack sprayer equipped with a 10-foot boom, fitted with six TJ-VS 8002 nozzles spaced 20 inches apart and directed forward and backward at a 45-degree angle. Fungicides were applied on May 22 and May 23 at the Feekes growth stage 10.5.1. All plots were inoculated with a mixture of isolates of *Fusarium graminearum* endemic to Indiana on May 23. The spore suspension (50,000 spores/ml) was applied at 300 ml/plot with the CO_2 backpack sprayer. Disease ratings were assessed on June 10. Fusarium head blight (FHB) incidence was measured as the number of infected heads out of 60 plants in each plot and calculated as a percentage. FHB severity was rated by visually assessing the percentage of the infected head, FHB index was calculated as (% FHB incidence multiplied by % FHB severity)/100 per plot. The eight center rows of each plot were harvested with a Kincaid plot combine on July 10, and yields were adjusted to 13.5% moisture. A subsample of grain was taken from each plot and partitioned for DON (deoxynivalenol) analysis completed by the University of Minnesota DON testing lab and to determine Fusarium damaged kernels (FDK) by visually assessing the percentage (0–100%) of the infected kernels. All data were analyzed in SAS 9.4 (SAS Institute, Cary, NC). A generalized linear mixed model analysis of variance was performed using PROC GLIMMIX. Values are least squares means, and values with different letters are significantly different based on the least squares difference test (α = 0.05).

In 2023, weather conditions were not favorable for FHB. No FHB or FDK was detected in the trial. Harvest moisture was highest when Miravis Ace was applied compared to the nontreated control (Table 17). The cultivar P25R61 had reduced harvest moisture, test weight, and yield as compared to P25R40. No significant differences were detected between treatments for test wight and yield.

TABLE 17. *Effect of Fungicide and Cultivar on Leaf Blotch, Deoxynivalenol (DON), Fusarium Damaged Kernels (FDK), and Yield of Wheat*

TREATMENT AND RATE/ACRE[z]	DON[y] PPM	FDK[x] %	HARVEST MOISTURE %	TEST WEIGHT LB/BU	YIELD[w] BU/ACRE
P25R40 (scab susceptible)	nd	0.0	16.3 a	54.2 a	88.5 a
P25R61 (scab resistant)	nd	0.0	16.2 b	52.9 b	80.6 b
Nontreated control	nd	0.0	16.0 c	53.6	85.8
Nontreated, noninoculated control	nd	0.0	16.0 c	53.3	79.7
Prosaro 421 SC 6.5 fl oz	nd	0.0	16.3 b	53.6	83.1
Miravis Ace 5.2 SC 13.7 fl oz	nd	0.0	16.6 a	53.8	83.9
Prosaro Pro 400 SC 10.3 fl oz	nd	0.0	16.3 b	53.7	87.3
Sphaerex 2.50 SC 7.3 fl oz	nd	0.0	16.3 b	53.4	87.5
P-value *cultivar*[v]	—	—	*0.0079*	*0.0001*	*0.0475*
P-value *treatment*	—	—	*0.0001*	*0.8363*	*0.8389*
P-value *cultivar*treatment*	—	—	*0.5668*	*0.9489*	*0.9564*

[z] Fungicide treatments applied on May 22 and May 23 at Feekes growth stage 10.5.1. All treatments contained a nonionic surfactant (Preference) at a rate of 0.125% v/v. All plots inoculated with *Fusarium graminearum* spore suspension (50,000 spores/ml) 24 hours after the treatment at Feekes 10.5.1. Spore suspension was applied at 300 ml/plot with the CO_2 handheld sprayer on May 23.

[y] Analysis of the mycotoxin deoxynivalenol (DON) was completed by the University of Minnesota DON Testing Lab on August 17. nd = not detected, DON <0.05 ppm.

[x] Fusarium damaged kernels (FDK) was visually assessed as a percentage (0–100%) of Fusarium damaged heads.

[w] Yields were adjusted to 13.5% moisture and harvested on July 10.

[v] All data were analyzed in SAS 9.4 (SAS Institute, Cary, NC). A generalized linear mixed model analysis of variance was performed using PROC GLIMMIX. Values are least squares means, and values with different letters are significantly different based on the least squares difference test (α = 0.05).

EVALUATION OF FOLIAR FUNGICIDES FOR WHEAT DISEASE MANAGEMENT IN CENTRAL INDIANA, 2021 (WHT23-07.ACRE)

E. A. Duncan, S. Shim, and D. E. P. Telenko, Department of Botany and Plant Pathology, Purdue University West Lafayette, IN 47907-2054

WHEAT (*TRITICUM AESTIVUM* P25R40)

Leaf blotch, *Septoria tritici/Stagnospora nodorum*

Plots were established at the Purdue Agronomy Center for Research and Education (ACRE) in Tippecanoe County, Indiana. The trial was a randomized complete block design with four replications. Plots were 7.5 feet wide and 20 feet long and consisted of 12 rows spaced 7.5 inches apart, and the center of each plot was used for evaluation. The previous crop was corn. On November 18, 2022, wheat cultivar P25R40 was drilled at 7.5 inches spacing. Foliar fungicides were applied on April 26 at Feekes growth stage 8 and on May 23 at Feekes 10.5.1 treatment. All fungicide applications were applied at 15 gal/acre and 40 psi using a CO_2 backpack sprayer equipped with a 10-foot boom, fitted with six TJ-VS 8002 nozzles spaced 20 inches apart. Foliar disease severity was rated by visually assessing the percentage of symptomatic leaf tissue on five flag leaves per plot for leaf blotch. Values for each plot were averaged before analysis. The eight center rows of each plot were harvested with a Kincaid small-plot combine on July 10, and yields were adjusted to 13.5% moisture. All data were analyzed in SAS 9.4 (SAS Institute, Cary, NC). A generalized linear mixed model analysis of variance was performed using PROC GLIMMIX. Values are least squares means, and values with different letters are significantly different based on the least squares difference test (α = 0.05).

In 2023, weather conditions were unfavorable for leaf blotch diseases. A low levels of leaf blotch was detected. Priaxor treatment had the lowest test weight when compared to the nontreated control and all other treatments except Tilt (Table 18). There was no significant effect on treatments for leaf blotch severity, harvest moisture, and yield of wheat.

TABLE 18. *Effect of Cultivar and Fungicide on Fusarium Head Blight (FHB), Deoxynivalenol (DON), Fusarium Damaged Kernels (FDK), and Yield of Wheat*

TREATMENT AND RATE/ACRE[z]	LEAF BLOTCH[y] %	HARVEST MOISTURE %	TEST WEIGHT LB/BU	YIELD[x] BU/ACRE
Nontreated control	0.0	13.8	55.3 ab	91.0
Nexicor Xemium 2.96 EC 7.0 fl oz at Feekes 8	0.1	13.9	56.3 a	93.5
Topguard 1.04 SC 10.0 fl oz at Feekes 8	0.0	13.7	55.0 ab	82.8
Priaxor 4.17 SC 4.0 fl oz at Feekes 8	0.1	13.6	53.5 c	77.9
Trivapro 2.21 SE 9.4 fl oz at Feekes 8	0.0	13.8	55.5 ab	83.1
Delaro 325 SC 8.0 fl oz at Feekes 8	0.0	14.0	55.9 ab	91.4
Quilt Xcel 2.2 SE 10.5 fl oz at Feekes 8	0.0	13.9	55.8 ab	89.2
Tilt 3.6 EC 4.0 fl oz at Feekes 8	0.2	13.7	54.4 bc	75.8
Headline 2.09 SC 6.0 fl oz at Feekes 8	0.1	13.8	55.5 ab	86.5
Adastrio 4.0 SC 6.0 fl oz at Feekes 8	0.1	13.7	55.4 ab	84.8
Prosaro 421 SC 6.5 fl oz at Feekes 10.5.1	0.1	14.0	55.5 ab	87.3
P-value[w]	0.3205	0.5689	0.0458	0.0559

[z] Fungicides were applied on April 26 at Feekes growth stage 8 and May 23 at the Feekes growth stage 10.5.1. All treatments contained a nonionic surfactant (Preference) at a rate of 0.125% v/v.

[y] Disease severity of leaf blotch was rated by visually assessing the percentage (0–100%) of symptomatic leaf tissue on five flag leaves per plot on June 10 and then averaged before analysis.

[x] Yields were adjusted to 13.5% moisture and harvested on July 10.

[w] All data were analyzed in SAS 9.4 (SAS Institute, Cary, NC). A generalized linear mixed model analysis of variance was performed using PROC GLIMMIX. Values are least squares means, and values with different letters are significantly different based on the least squares difference test (α = 0.05).

PINNEY PURDUE AGRICULTURAL CENTER (PPAC)

UNIFORM FUNGICIDE COMPARISON FOR TAR SPOT IN CORN IN NORTHWESTERN INDIANA, 2023 (COR23-02.PPAC)

M. S. Mizuno, S. Shim, and D. E. P. Telenko, Department of Botany and Plant Pathology, Purdue University West Lafayette, IN 47907-2054

CORN (*ZEA MAYS* W2585VT2PRIB)

Tar spot, *Phyllachora maydis*

A trial was established at the Pinney Purdue Agricultural Center (PPAC) in Porter County, Indiana. The experiment was a randomized complete block design with four replications. Plots were 10 feet wide and 30 feet long and consisted of four rows, and the two center rows were used for evaluation. The previous crop was corn. Standard practices for grain corn production in Indiana were followed. Corn hybrid W2585VT2PRIB was planted in 30-inch row spacing at a rate of 34,000 seeds/acre on May 22. The field was overhead irrigated weekly at 1 inch. unless weekly rainfall was 1 inch or higher to encourage disease. All foliar fungicide applications were applied at 15 gal/acre and at 40 psi using a Lee self-propelled sprayer equipped with a 10-foot boom, fitted with six TJ-VS 8002 nozzles spaced 20 inches apart. Fungicides were applied on August 3 at blister (R2) growth stage and three weeks after treatments on August 22 at dough (R4) growth stage. Disease ratings were assessed on September 6, September 19, and October 16 at dough (R4), early dent (R5), and late dent (R5) growth stages, respectively. Tar spot was rated by visually assessing the percentage of stromata per leaf (0–100%) on five plants in each plot at the ear leaf. Percent of canopy green was rated by visually assessing the percentage (0–100%) of the whole plot for crop canopy that remained green at dent (R5) growth stage. The two center rows of each plot were harvested on November 7, and yields were adjusted to 15.5% moisture. All disease and yield data were analyzed in SAS 9.4 (SAS Institute, Cary, NC). A generalized linear mixed model analysis of variance was performed using PROC GLIMMIX. Values are least squares means, and values with different letters are significantly different based on the least squares difference test (α = 0.05).

In 2023, weather conditions were favorable for disease. Tar spot was the most prominent disease in the trial and reached moderate severity. All treatments significantly reduced tar spot severity compared to the nontreated control on September 6 and September 19 (Table 19). On September 6, Veltyma followed by Headline

AMP and Headline AMP followed by Aproach Prima were the treatments that showed the greatest reduction in the severity of tar spot stromata but were not significantly different from Delaro Complete, Aproach Prima followed by Headline AMP, Miravis Neo followed by Headline AMP, and Headline AMP followed by Veltyma, Delaro Complete, or Headline AMP. On September 19, no significant difference was detected between fungicide programs. On October 16, only Veltyma followed by Headline AMP, Aproach Prima followed by Headline AMP, Miravis Neo followed by Headline AMP, Headline AMP followed by Veltyma, and Headline AMP followed by Aproach Prima reduced tar spot severity over the nontreated control. Veltyma followed by Headline AMP, Aproach Prima followed by Headline AMP, and Headline APM followed by Veltyma programs significantly increased canopy greenness over the nontreated control. There was no significant effect of treatment on yield of corn.

TABLE 19. *Effect of Fungicide Programs on Tar Spot Severity, Canopy Greenness, and Yield of Corn*

TREATMENT, RATE/ACRE, AND TIMING[z]	TAR SPOT %[y] SEPTEMBER 6	TAR SPOT %[y] SEPTEMBER 19	TAR SPOT %[y] OCTOBER 16	CANOPY GREEN[x] %	YIELD[w] BU/ACRE
Nontreated control	2.3 a	12.8 a	19.6 a	0.3 d	194.2
Veltyma 3.34 S 7 fl oz at R2	0.9 b-e	3.0 b	17.1 abc	1.5 d	213.6
Aproach Prima 2.34 SC 6.8 fl oz at R2	1.2 b	3.8 b	17.6 ab	2.8 d	186.6
Miravis Neo 2.5 SE 13.7 fl oz at R2	1.0 bc	4.8 b	18.9 ab	3.8 d	193.6
Delaro Complete 458 SC 8 fl oz at R2	0.7 c-f	1.7 b	16.0 a-d	15.0 bcd	220.7
Headline AMP 1.68 SC 10 fl oz at R2	1.1 bc	3.3 b	16.5 a-d	2.5 d	194.4
Veltyma 3.34 S 7 fl oz at R1 fb					
Headline AMP 1.68 SC 10 fl oz at 3 WAT	0.4 f	1.1 b	8.2 f	31.3 a	221.0
Aproach Prima 2.34 SC 6.8 fl oz at R2 fb					
Headline AMP 1.68 SC 10 fl oz at 3 WAT	0.5 def	2.5 b	11.8 def	22.5 abc	214.5
Miravis Neo 2.5 SE 13.7 fl oz at R2 fb					
Headline AMP 1.68 SC 10 fl oz at 3 WAT	0.7 c-f	2.5 b	12.2 c-f	6.5 d	211.2
Delaro Complete 458 SC 8 fl oz at R2 fb					
Headline AMP 1.68 SC 10 fl oz at 3 WAT	1.0 bcd	3.9 b	16.0 a-d	3.8 d	216.9
Headline AMP 1.68 SC 10 fl oz at R2 fb					
Veltyma 3.34 S 7 fl oz at 3 WAT	0.6 c-f	1.1 b	10.1 ef	27.5 ab	230.9
Headline AMP 1.68 SC 10 fl oz at R2 fb					
Aproach Prima 2.34 SC 6.8 fl oz at 3 WAT	0.4 efç	0.8 b	14.4 b-e	12.5 bcd	185.6
Headline AMP 1.68 SC 10 fl oz at R2 fb					
Miravis Neo 2.5 SE 13.7 fl oz at 3 WAT	0.9 b-e	3.0 b	14.7 a-e	1.5 d	212.5
Headline AMP 1.68 SC 10 fl oz at R2 fb					
Delaro Complete 458 SC 8 fl oz at 3 WAT	0.8 b-f	2.0 b	14.7 a-e	14.5 bcd	191.3
Headline AMP 1.68 SC 10 fl oz at R2 fb					
Headline AMP 1.68 SC 10 fl oz at 3 WAT	0.8 b-f	2.4 b	14.8 a-e	11.5 cd	192.5
P-value[v]	*0.0001*	*0.0001*	*0.0015*	*0.0006*	*0.1607*

[z] Fungicides were applied on August 3 at blister (R2) and on August 22, three weeks after treatment at dough (R4) growth stage. All treatments applied contained a nonionic surfactant (Preference) at a rate of 0.25% v/v. fb = followed by, WAT = weeks after treatment.

[y] Tar spot severity was visually assessed as a percentage (0–100%) of leaf area on five plants in each plot at the ear leaf on September 6, September 19, and October 16 at dough (R4) and early dent (R5) and late dent (R5) growth stages, respectively.

[x] Canopy greenness was visually assessed as a percentage (0–100%) of crop canopy green on October 16 at late dent (R5) growth stage.

[w] Yields were adjusted to 15.5% moisture and harvested on November 7.

[v] All data were analyzed in SAS 9.4 (SAS Institute, Cary, NC). A generalized linear mixed model analysis of variance was performed using PROC GLIMMIX. Values are least squares means, and values with different letters are significantly different based on the least squares difference test (α = 0.05).

EVALUATION OF HYBRID AND FUNGICIDE TIMING FOR TAR SPOT IN CORN IN NORTHWESTERN INDIANA, 2023 (COR23-03.PPAC)

K. M. Goodnight, S. Shim, and D. E. P. Telenko, Department of Botany and Plant Pathology, Purdue University West Lafayette, IN 47907-2054

CORN (*ZEA MAYS* W2585VT20 AND P0589AMXT)

Tar spot, *Phyllachora maydis*

A trial was established at the Pinney Purdue Agricultural Center (PPAC) in Porter County, Indiana. The experiment was a randomized complete block design with four replications. Plots were 10 feet wide and 30 feet long and consisted of four rows, and the two center rows were used for evaluation. The previous crop was corn. Standard practices for grain corn production in Indiana were followed. Corn hybrids W2585VT-2PRIB (tar spot susceptible) and P0589AMXT (tar spot resistant) were planted in 30-inch row spacing at a rate of 34,000 seeds/acre on May 18. All fungicide applications were applied at 15 gal/acre and at 40 psi using a Lee self-propelled sprayer equipped with a 10-foot boom, fitted with six TJ-VS 8002 nozzles spaced 20 inches apart. Delaro Complete 458 SC at 8 fl oz/acre fungicide was applied on July 25, August 3, August 22, and August 29 at the 10-leaf (V10), tassel/silk (VT/R1), blister (R2), and dough (R4) growth stages, respectively. A weather-based prediction model, Tarspotter (https://ipcm.wisc.edu/apps/tarspotter/) was used, and applications were made on August 17 and August 29 at the blister (R2) and dough (R4) growth stages, respectively. Disease ratings were assessed on September 21 at dent (R5) and on October 10 at maturity (R6) growth stage. Tar spot was rated by visually assessing the percentage of stromata (0–100%) per leaf on five plants in each plot at the ear leaf. Values for the five leaves were averaged before analysis. Percentage canopy greenness was rated by visually assessing the percentage (0–100%) of the whole plot crop canopy that remained green at maturity (R6) growth stage. The two center rows of each plot were harvested on November 6, and yields were adjusted to 15.5% moisture. All disease and yield data were analyzed in SAS 9.4 (SAS Institute, Cary, NC). A generalized linear mixed model analysis of variance was performed using PROC GLIMMIX. Values are least squares means, and values with different letters are significantly different based on the least squares difference test (α = 0.05).

In 2023, weather conditions were favorable for disease. Tar spot was the most prominent disease in the trial and reached moderate severity. There was significant interaction between hybrid and fungicide; therefore, treatment effect is evaluated across each hybrid (Table 20). For the tar spot-susceptible hybrid, all treatments significantly reduced tar spot stromata compared to the nontreated control for all three disease ratings. For the hybrid moderately resistant to tar spot, all treatments applied except for 10th-leaf (V10) for the September 7 and September 21 ratings and treatments applied at 10th-leaf (V10), tassel/silk (VT/R1), and dough (R4) growth stage for October 10 rating significantly reduced tar spot stromata compared to the nontreated control. No significant treatment differences were detected for canopy greenness for either hybrid. Only the Tarspotter application significantly increased yield compared to the nontreated control for the tar spot-susceptible hybrid. There was no significant effect on corn yield for the hybrid moderately resistant to tar spot.

TABLE 20. *Effect of Fungicide on Tar Spot Severity, Canopy Greenness, and Yield of Corn*

TREATMENT, RATE/ACRE, AND TIMING[z]	TAR SPOT[y] % SEPTEMBER 21		TAR SPOT[y] % OCTOBER 10		CANOPY GREEN[x] %		TEST WEIGHT LB/BU		YIELD[w] BU/ACRE	
	TS[v]	TMR[v]	TS	TMR	TS	TMR	TS	TMR	TS	TMR
Nontreated control	1.1 a	0.5 a	2.2 a	0.6 a	11.5 a	5.4 a	37.5	31.3	212.6 b	211.7
Delaro Complete 458 SC at V10	0.5 bc	0.3 ab	1.1 b	0.5 ab	7.5 b	4.8 a	43.8	53.8	230.5 b	212.7
Delaro Complete 458 SC at VT/R1	0.3 bc	0.2 bc	0.6 bc	0.3 b	5.4 b	4.3 ab	61.3	43.8	226.4 b	209.0
Delaro Complete 458 SC at R2	0.5 b	0.2 bc	1.0 b	0.4 b	4.7 bc	1.7 bc	56.3	52.5	226.3 b	225.9
Delaro Complete 458 SC at R4	0.5 b	0.3 bc	1.2 b	0.3 bc	5.1 bc	2.8 abc	60.0	55.0	218.5 b	223.4
Delaro Complete 458 SC based on Tarspotter (R2 fb R4)	0.2 c	0.2 c	0.2 c	0.1 c	2.2 c	1.0 c	68.8	55.0	256.4 a	235.2
P-value[u]	*0.0003*	*0.0082*	*0.0002*	*0.0027*	*0.0003*	*0.0144*	*0.0589*	*0.1437*	*0.0062*	*0.2025*

[z] Fungicide treatments were applied on July 25, August 3, August 17, August 22, and August 29 at 10th-leaf (V10), tassel/silk (VT/R1), blister (R2), and dough (R4) growth stages, respectively. Tarspotter applications were made on August 17 and August 29 at the blister (R2) and dough (R4) growth stages, respectively. fb = followed by.

[y] Tar spot stromata was visually assessed as the percentage (0–100%) of affected leaf area on five plants in each plot at the ear leaf on September 21 and October 10.

[x] Canopy greenness was visually assessed as a percentage (0–100%) green of the plot as a whole on October 10.

[w] Yields were adjusted to 15.5% moisture and harvested on November 6.

[v] TS = tar spot susceptible (W2585VT2PRIB), TMR = tar spot moderately resistant (P0589AMXT).

[u] All data were analyzed in SAS 9.4 (SAS Institute, Cary, NC). A generalized linear mixed model analysis of variance was performed using PROC GLIMMIX. Values are least squares means, and values with different letters are significantly different based on the least squares difference test (α = 0.05).

EVALUATION OF TAR SPOT MANAGEMENT PROGRAMS IN ORGANIC CORN IN NORTHWESTERN INDIANA, 2023 (COR23-04.PPAC)

C. Rocco da Silva, S. Shim, and D. E. P. Telenko, Department of Botany and Plant Pathology, Purdue University West Lafayette, IN 47907-2054

CORN (*ZEA MAYS* 0.52-96 AND 0.51-95)

Tar spot, *Phyllachora maydis*

A trial was established at the Pinney Purdue Agricultural Center (PPAC) in Porter County, Indiana. The experiment was a randomized complete block design with four replications. Plots were 10 feet wide and 30 feet long and consisted of four rows, and the two center rows were used for evaluation. The previous crop was corn. Standard practices for grain corn production in Indiana were followed. Corn organic hybrids 0.52-6 and 0.51-95 were planted in 30-inch row spacing at a rate of 34,000 seeds/acre on May 18. The field was overhead irrigated weekly at 1 inch, unless weekly rainfall was 1 inch or higher, to encourage disease. All fungicide applications were applied at 15 gal/acre and at 40 psi using a Lee self-propelled sprayer equipped with a 10-foot boom, fitted with six TJ-VS 8002 nozzles spaced 20 inches apart at 3.6 mph. Fungicide treatments were applied on August 3 at silk (R1) growth stage. Disease ratings were assessed on September 9 and September 15 at dough (R4) and dent (R5) growth stages, respectively. Tar spot was rated by visually assessing the percentage of stromata (0–100%) per leaf on five plants in each plot at the ear leaf. Values for the five leaves were averaged before analysis. Percent canopy of greenness was rated by visually assessing the percentage (0–100%) of the whole plot for crop canopy that remained green at dent (R5) growth stage. The two center rows of each plot were harvested on November 7, and yields were adjusted to 15.5% moisture. All disease and yield data were analyzed in SAS 9.4 (SAS Institute, Cary, NC). A generalized linear mixed model analysis of variance was performed using PROC GLIMMIX. Values are least squares means, and values with different letters are significantly different based on the least squares difference test (α = 0.05).

In 2023, weather conditions were favorable for disease. Tar spot was the most prominent disease in the trial and reached moderate severity. There was no significant interaction between hybrid and fungicide for disease and yield; therefore, main effects of hybrid and fungicide were evaluated. No differences between hybrids were observed for tar spot severity and yield (Table 21). Tar spot severity was significantly reduced over the nontreated control by Headline AMP and Badge X2 on September 9 and September 15 but were not significantly different from Serifel and Actinovate on September 15. The percentage of canopy green was highest in the hybrid 0.51-95 and when treated with Headline AMP but not significantly different from Badge X2. There were no significant differences in treatments and nontreated control for grain yield.

TABLE 21. *Effect of Fungicide on Tar Spot Severity, Canopy Greenness, and Yield of Corn*

TREATMENT AND RATE/ACRE[z]	TAR SPOT[y] % SEPTEMBER 9	TAR SPOT[y] % SEPTEMBER 15	CANOPY GREEN[x] %	YIELD[w] BU/ACRE
Hybrids				
0.52-96	3.1	10.7	53.1 b	198.5
0.51-95	2.9	10.7	60.6 a	197.7
Fungicide programs				
Nontreated control	4.2 a	18.1 a	51.3 b	199.1
Headline AMP 1.68 SE 10 fl oz	0.9 b	2.4 b	68.8 a	204.3
Serifel WP 16 fl oz	3.7 a	10.6 ab	51.3 b	195.8
Actinovate AG 12 ox	3.7 a	11.1 ab	57.5 b	198.9
Badge X2 SC 1.8 lb	1.3 b	6.0 b	60.0 ab	193.3
OxiDate 5.0 128 fl oz	4.3 a	15.8 a	52.5 b	197.2
P-value *hybrid*[v]	0.0061	0.9967	0.0187	0.2518
P-value *fungicide*	0.7740	0.0147	0.0141	0.7756
P-value *hybrid*fungicide*	0.1068	0.0859	0.0476	0.1061

[z] Fungicide treatments were applied at on August 3 at silk (R1) growth stage.

[y] Tar spot stromata severity was visually assessed as a percentage (0–100%) of leaf area on five plants in each plot at the ear leaf on September 9 and September 15 at dough (R4) and dent (R5) growth stages, respectively.

[x] Canopy greenness was visually assessed as percentage (0–100%) of green of the plot as a whole on September 15.

[w] Yields were adjusted to 15.5% moisture and harvested on November 7.

[v] All data were analyzed in SAS 9.4 (SAS Institute, Cary, NC). A generalized linear mixed model analysis of variance was performed using PROC GLIMMIX. Values are least squares means, and values with different letters are significantly different based on the least squares difference test (α = 0.05).

EVALUATION OF TILLAGE, HYBRID, AND FUNGICIDE EFFICACY FOR DISEASES IN CORN IN NORTHWESTERN INDIANA, 2023 (COR23-05.PPAC)

S. Shim and D. E. P. Telenko, Department of Botany and Plant Pathology, Purdue University West Lafayette, IN 47907-2054

CORN (*ZEA MAYS* W2585SSRIB, P0589AMXT)

Tar spot, *Phyllachora maydis*
Gray leaf spot, *Cercospora zeae-maydis*
Northern corn leaf blight, *Exserohilum turcicum*

A trial was established at the Pinney Purdue Agricultural Center (PPAC) in Porter County, Indiana. The trial was a randomized complete block design with six replications. Plots were 10 feet wide and 30 feet long and consisted of four rows, and the two center rows were used for evaluation. The previous crop was corn. Standard practices for nonirrigated grain corn production in Indiana were followed. Corn hybrid W2585SSRIB (tar spot susceptible) and P0589AMXT (tar spot moderate resistant) were planted in 30-inch row spacing at a rate of 34,000 seeds/foot on May 18. Veltyma application was applied at 15 gal/acre and at 40 psi using a Lee self-propelled sprayer equipped with a 10-foot boom, fitted with six TJ-VS 8002 nozzles spaced 20 inches apart. Fungicide was applied on August 2 at R2 (full bloom). Tar spot was rated by visually assessing the percentage of stroma per leaf on 10 plants in each plot at the ear leaf on September 28 at maturity (R6) growth stage. Gray leaf spot (GLS) and northern corn leaf blight (NCLB) were rated by visually assessing the percentage severity on ear leaf on 10 plants on September 7 at dough/dent (R4/R5) growth stage. Values for each plot were averaged before analysis. The two center rows of each plot were harvested on November 7, and yields were adjusted to 15.5% moisture. All data were analyzed in SAS 9.4 (SAS Institute, Cary, NC). A generalized linear mixed model analysis of variance was performed using PROC GLIMMIX. Values are least squares means, and values with different letters are significantly different based on the least squares difference test ($\alpha = 0.05$).

In 2023, weather conditions were favorable for disease. Tar spot was the most prominent disease in the trial and reached moderate severity. There was no significant interaction between tillage and hybrid and fungicide; therefore, main effects of tillage are presented. There was a significant interaction between hybrid and fungicide for tar spot severity; therefore, data presented for that interaction. No significant differences were detected between no tillage and tillage main effects for tar spot, GLS, NCLB, and grain yield (Table 22). Harvest moisture was reduced under tillage versus no tillage, while test weight was highest in the tillage verses no-tillage treatments. Tar spot severity was highest in the susceptible hybrid with no fungicide application. Veltyma fungicide application significantly reduced tar spot severity compared to the nontreated susceptible hybrid, but no significant differences were detected in the moderately resistant hybrid. There were no significant differences between hybrid and fungicide programs and GLS and NCLB severity, harvest moisture, test weight, and yield.

TABLE 22. *Effect of Tillage, Hybrid, and Fungicide for Foliar Diseases in Corn and Yield of Corn*

TILLAGE, HYBRID, TREATMENT, AND TIMING[z]	TAR SPOT[y] %	GLS[x] %	NCLB[x] %	HARVEST MOISTURE %	TEST WEIGHT LB/BU	YIELD[w] BU/ACRE
No tillage (high residue)	12.9	0.16	0.03	22.4 a	47.4 b	214.3
Tillage (low residue)	12.2	0.03	0.12	20.1 b	50.1 a	229.9
Susceptible, Nontreated control	23.1 a	0.06	0.14	21.4	46.9	215.5
Susceptible, Veltyma 7.0 fl oz/acre	10.0 b	0.02	0.16	22.7	53.2	237.8
Moderately resistant, Nontreated control	10.4 b	0.00	0.00	20.0	46.3	211.8
Moderatey resistant, Veltyma 7.0 fl oz/acre	6.8 b	0.31	0.01	20.7	48.6	223.4
P-value *tillage*[v]	*0.7464*	*0.2661*	*0.4989*	*0.0003*	*0.0060*	*0.0004*
P-value *hybrid*	*0.0024*	*0.3019*	*0.3822*	*0.0090*	*0.0066*	*0.0199*
P-value *fungicide*	*0.0013*	*0.2152*	*0.8756*	*0.0277*	*0.0001*	*0.0002*
P-value *tillage*hybrid*	*0.7030*	*0.2439*	*0.3822*	*0.4437*	*0.7392*	*0.3051*
P-value *tillage*fungicide*	*0.1141*	*0.4880*	*0.8756*	*0.1958*	*0.0723*	*0.1692*
P-value *hybrid*fungicide*	***0.0453***	*0.1802*	*0.9675*	*0.5000*	***0.0342***	*0.1570*
P-value *tillage*hybrid*fungicide*	*0.6830*	*0.3101*	*0.9675*	*0.1845*	*0.7922*	*0.5632*

[z] Veltyma application was applied at 15 gal/acre and 40 psi using a Lee self-propelled sprayer equipped with a 10-foot boom, fitted with six TJ-VS 8002 nozzles spaced 20 inches apart. Veltyma was applied on August 2 at blister (R2).

[y] Tar spot stroma visually assessed as a percentage (0–100%) of ear leaf on 10 plants in each plot on September 28 at maturity (R6) growth stage.

[x] GLS and NCLB severity were visually assessed as a percentage (0–100%) of leaf area on 10 plants in each plot on September 7 at dough/dent (R4/R5) growth stage. GLS = gray leaf spot, NCLB = northern corn leaf blight.

[w] Yields were adjusted to 15.5% moisture and harvested on November 7.

[v] All data were analyzed in SAS 9.4 (SAS Institute, Cary, NC). A generalized linear mixed model analysis of variance was performed using PROC GLIMMIX. Values are least squares means, and values with different letters are significantly different based on the least squares difference test (α = 0.05).

COMPARISON OF HYBRID MATURITY, PLANTING, AND FUNGICIDE FOR TAR SPOT IN CORN (COR23-07.PPAC)

J. D. Peña, S. Shim, and D. E. P. Telenko, Department of Botany and Plant Pathology, Purdue University West Lafayette, IN 47907-2054

CORN (*ZEA MAYS* P9608Q AND P1099Q)

Tar spot, *Phyllachora maydis*

A trial was established at the Pinney Purdue Ag Center (PPAC) in Porter County, Indiana. The experiment was a randomized complete block design with four replications. Plots were 10 feet wide and 30 feet long and consisted of four rows, and the two center rows were used for evaluation. The previous crop was corn. Standard practices for nonirrigated grain corn production in Indiana were followed. Corn hybrids P9608Q and P1099Q were planted in 30-inch row spacing at a rate of 34,000 seeds/acre on May 18 and May 26. P9608Q is a 96-day hybrid, and P1099Q is a 110-day hybrid. Foliar fungicide applications were made at the tassel/silk (VT/R1) growth stage on August 3 for the May 18 and on August 10 for the May 26 planting plots. All foliar fungicide applications were applied at 15 gal/acre and at 40 psi using a Lee self-propelled sprayer equipped with a 10-foot boom, fitted with six TJ-VS 8002 nozzles spaced 20 inches apart. Disease ratings were assessed on September 20 and October 10 at dough (R4) and physiological maturity (R6), respectively. Tar spot severity was visually assessed as a percentage (0–100%) of symptomatic leaf area at ear leaf on five plants per plot and averaged before analysis. The two center rows of each plot were harvested on November 6, and yields were adjusted to 15.5% moisture. All disease and yield data were analyzed in SAS 9.4 (SAS Institute, Cary, NC). A generalized linear mixed model analysis of variance was performed using PROC GLIMMIX. Values are least squares means, and values with different letters are significantly different based on the least squares difference test ($\alpha = 0.05$).

In 2023, weather conditions were favorable for disease. Tar spot was the most prominent disease in the trial and reached moderate severity. On September 20, hybrid P9608Q had significantly more tar spot than P1099Q, and a Veltyma fungicide application significantly reduced the severity of tar spot stromata in P9608Q but not P1099Q (Table 23). On October 10 severity of tar spot was significantly reduced in Hybrid P1099Q compared to Hybrid P9608Q and with Veltyma 3.34 S compared to the nontreated control. Canopy greenness on October 3 was significantly higher at planting date B over planting date A in hybrid P1099Q as compared to P9608Q and increased with Veltyma over nontreated. Yield was highest for Hybrid P9608Q planted on May 26 (planting date B) verses at the planting date of May 18 (planting date B) but was not significantly more than P1099Q at the same planting dates.

TABLE 23. *Effect of Treatments on Foliar Disease Severity in Corn and Yield of Corn*

TREATMENT AND RATE/ ACRE[z]	TAR SPOT[y] % SEPTEMBER 20		TAR SPOT[y] % OCTOBER 10	CANOPY GREEN[x] %	HARVEST MOISTURE %	TEST WEIGHT LB/BU	YIELD[w] BU/ACRE	
Planting Date A	2.4		3.0	42.2 a	22.9 a	53.2		
Planting Date B	2.0		3.0	57.8 b	24.8 b	52.1		
							PD A	PD B
Hybrid P9608Q			3.7 a	23.4 a	20.0 a	54.6 a	182.6 c	222.5 a
Hybrid P1099Q			2.2 b	76.6 b	27.6 b	50.7 b	197.8 bc	204.2 ab
	P9608Q	P1099Q						
Nontreated control	6.4 a	0.8 b	4.2 a	45.6 a	23.5	52.7	199.6	
Veltyma 3.34 S 7.0 fl oz	1.2 b	0.3 b	1.8 b	54.4 b	24.1	52.6	204.0	
P-value *date*[v]	0.6233		0.9979	0.0001	0.0003	0.0600	0.8042	
P-value *hybrid*	0.0002		0.0039	0.0001	0.0001	0.0001	0.0013	
P-value *fungicide*	0.0006		0.0001	0.0066	0.1590	0.8202	0.4849	
P-value *planting date*hybrid*	0.9378		0.8739	0.6708	0.0627	0.6500	0.0142	
P-value *planting date*fungicide*	0.9630		0.7289	0.8314	0.4090	0.8686	0.8213	
P-value *hybrid*fungicide*	0.0034		0.1056	0.0659	0.8029	0.6798	0.3237	
P-value *planting date*hybrid*fungicide*	0.8490		0.1802	0.0429	0.4903	0.6063	0.2555	

[z] Foliar applications were made at the tassel/silk (VT/R1) growth stage on August 3 for May 18 planting plots and August 10 for May 26 planting plots.

[y] Tar spot stromata was visually assessed as a percentage (0–100%) of leaf area on five plants in each plot at the ear leaf on September 20 and October 10.

[x] Canopy greenness was visually assessed as a percentage (0–100%) of crop canopy green on October 3.

[w] Yields were adjusted to 15.5% moisture and harvested on November 6. PD = planting date.

[v] All data were analyzed in SAS 9.4 (SAS Institute, Cary, NC). A generalized linear mixed model analysis of variance was performed using PROC GLIMMIX. Values are least squares means, and values with different letters are significantly different based on the least squares difference test (α = 0.05).

EVALUATION OF FUNGICIDES FOR FOLIAR DISEASES IN CORN IN NORTHWESTERN INDIANA, 2023 (COR23-13.PPAC)

E. A. Duncan and D. E. P. Telenko, Department of Botany and Plant Pathology, Purdue University West Lafayette, IN 47907-2054

CORN (*ZEA MAYS* W2585VT2PRIB)

Tar spot, *Phyllachora maydis*

A trial was established at the Pinney Purdue Agricultural Center (PPAC) in Porter County, Indiana. The experiment was a randomized complete block design with four replications. Plots were 10 feet wide and 30 feet long and consisted of four rows, and the two center rows were used for evaluation. The previous crop was soybean. Standard practices for nonirrigated grain corn production in Indiana were followed. Corn hybrid W2585VT2PRIB was planted in 30-inch row spacing at a rate of 34,000 seeds/acre on May 25. Foliar applications were made at V6/V7, V10, and blister (R2) growth stages on July 5, July 17, and August 2, respectively. All foliar fungicide applications were applied at 15 gal/acre and 40 psi using a Lee self-propelled sprayer equipped with a 10-foot boom, fitted with six TJ-VS 8002 nozzles spaced 20 inches apart. Disease ratings were assessed on August 28, September 8, and October 3 at dough (R4), dent (R5), and physiological maturity (R6) growth stages, respectively. Tar spot severity was visually assessed as a percentage (0–100%) of symptomatic leaf area at ear leaf on five plants per plot and averaged before analysis. The two center rows of each plot were harvested on November 3, and yields were adjusted to 15.5% moisture. All data were analyzed in SAS 9.4 (SAS Institute, Cary, NC). A generalized linear mixed model analysis of variance was performed using PROC GLIMMIX. Values are least squares means, and values with different letters are significantly different based on the least squares difference test ($\alpha = 0.05$).

In 2023, weather conditions were favorable for tar spot. Tar spot was the most prominent disease in the trial and reached a moderate level. All fungicide programs significantly reduced tar spot on the ear leaf on August 28 over the nontreated control except Lucento at R2 and ALB4003B 14.0 oz + ALB5018 0.10% at V6 followed by ALB4003B 14.0 oz + ALB5018 0.10% at R2 (Table 24). All fungicide programs reduced tar spot on ear leaves on September 8 over the nontreated control. There was no significant difference between treatments and the nontreated control for tar spot on October 3. Trivapro, Delaro Complete, Veltyma, and ALB4003B 14.0 oz + ALB5018 0.10% at V6 followed by ALB4016 14.4 oz + ALB5018 0.10% + ALB4021B 32.0 oz at R2 had significantly higher canopy greenness on October 3 as compared to the nontreated control. Adastrio had the highest harvest moisture but was not significant from the nontreated control. There were no significant differences between treatments and the nontreated control for test weight and yield of corn.

TABLE 24. *Effect of Fungicide on Tar Spot Severity, Canopy Greenness, and Yield of Corn*

TREATMENT, RATE/ACRE, AND TIMING[z]	TAR SPOT[y] % AUGUST 28	TAR SPOT[y] % SEPTEMBER 8	TAR SPOT[y] % OCTOBER 3	CANOPY GREEN[x] %	HARVEST MOISTURE %	TEST WEIGHT LB/BU	YIELD[w] BU/ACRE
Nontreated control	0.4 a	1.8 a	27.5	32.5 e	22.4 a-d	52.5	235.8
Trivapro 2.21 SE 13.7 fl oz at R2	0.2 d	0.8 bcd	25.5	55.0 abc	23.2 a-d	52.0	243.0
Delaro Complete 458 SC 8.0 fl oz at R2	0.1 d	0.4 cd	20.4	70.0 a	22.9 abc	51.7	243.0
Veltyma 3.34 S 7.0 fl oz at R2	0.2 bcd	0.2 d	21.5	60.0 ab	23.0 ab	51.1	243.5
Lucento 7.17 SC 5.0 fl oz at R2	0.4 abc	0.9 bc	26.0	32.5 e	21.7 d	52.8	232.6
Adastrio 4.0 SC 8.0 fl oz at R2	0.2 cd	0.5 bcd	21.8	47.5 b-e	23.1 a	51.7	241.7
Topguard EQ 4.29 SC 5.0 fl oz at R2	0.2 d	0.5 bcd	18.0	42.5 cde	22.1 cd	52.4	240.2
Topguard 1.04 SC 10.0 fl oz at V6/V7 fb Adastrio 4.0 SC 8.0 fl oz at R2	0.4 ab	0.6 bcd	24.5	43.8 b-e	21.8 d	51.8	243.4
OR-009EPA 0.40 % v/v + Delaro Complete 458 SC 8.0 fl oz at R2	0.1 d	0.3 cd	19.8	47.5 b-e	22.5 a-d	55.4	245.1
OR-009EPA 0.40 % v/v + Trivapro 2.21 SE 13.7 fl oz at R2	0.2 bcd	1.0 bc	26.0	46.3 b-e	22.8 abc	51.7	242.3
ALB4003B 14.0 oz + ALB5018 0.10% at V6 fb ALB4016 14.4 oz + ALB5018 0.10% + ALB4021B 32.0 oz at R2	0.2 d	0.4 cd	20.8	52.5 bcd	22.2 bcd	52.4	243.9
ALB4003B 14.0 oz + ALB5018 0.10% at V6 fb ALB4003B 14.0 oz + ALB5018 0.10% at R2	0.4 ab	1.1 b	26.5	36.3 de	21.7 d	53.2	240.8
P-value[v]	*0.0006*	*0.0013*	*0.0746*	*0.0013*	*0.0088*	*0.5528*	*0.8920*

[z] Fungicide treatments were applied at V6/V7, V10, and blister (R2) growth stages on July 5, July 17, and August 2, respectively. All foliar fungicide applications were applied at 15 gal/acre. fb = followed by.

[y] Tar spot stromata was visually assessed as a percentage (0–100%) of leaf area on five plants in each plot at the ear leaf on August 28, September 8, and October 3 at dough (R4), dent (R5), and physiological maturity (R6) growth stages, respectively.

[x] Canopy greenness was visually assessed as a percentage (0–100%) of crop canopy on October 3.

[w] Yields were adjusted to 15.5% moisture and harvested on November 3.

[v] All data were analyzed in SAS 9.4 (SAS Institute, Cary, NC). A generalized linear mixed model analysis of variance was performed using PROC GLIMMIX. Values are least squares means, and values with different letters are significantly different based on the least squares difference test (α = 0.05).

EVALUATION OF DRONE APPLICATIONS FOR TAR SPOT IN CORN IN NORTHWESTERN INDIANA, 2023 (COR23-16.PPAC)

M. S. Mizuno, S. Shim, and D. E. P. Telenko, Department of Botany and Plant Pathology, Purdue University West Lafayette, IN 47907-2054

CORN (*ZEA MAYS* W2585VT2PRIB)

Tar spot, *Phyllachora maydis*

A trial was established at the Pinney Purdue Agricultural Center (PPAC) in Porter County, Indiana. The experiment was a randomized complete block design with four replications. Plots were 10 feet wide and 30 feet long and consisted of four rows, and the two center rows were used for evaluation. The previous crop was corn. Standard practices for grain corn production in Indiana were followed. Corn hybrid W2585VT2PRIB was planted in 30-inch row spacing at a rate of 34,000 seeds/acre on May 22. The field was overhead irrigated weekly at 1 inch unless weekly rainfall was 1 inch or higher to encourage disease. Veltyma 3.34 S 7 fl oz/acre was applied on August 21 at silk (R1) growth stage using three different applicators: a Lee self-propelled sprayer equipped with a 10-foot boom, fitted with six TJ-VS 8002 nozzles spaced 20 inches apart at 3.6 mph; a CO_2 backpack sprayer equipped with a 10-foot boom, fitted with four TJ-VS 8002 nozzles spaced 20 inches apart at 3.1 mph, both applied at 15 gal/acre and 40 psi; and a DJI Agras T10 drone equipped with a 2.1-gal spray tank with spray pattern using four TJ-VS 8002 nozzles spaced 20 inches apart at 3.1 mph and applied at 1.65 gal/acre and 40 psi. Disease ratings were assessed on September 6, September 29, and October 16 at early dent (R5) and late dent (R5) growth stages, respectively. Tar spot stroma severity was rated by visually assessing the percentage (0–100%) per leaf on five plants in each plot at the ear leaf. Values for the five leaves were averaged before analysis. Percent of canopy greenness was rated by visually assessing the percentage (0–100%) of the whole plot for crop canopy that remained green at late dent (R5) growth stage. The two center rows of each plot were harvested on November 3, and yields were adjusted to 15.5% moisture. All disease and yield data were analyzed in SAS 9.4 (SAS Institute, Cary, NC). A generalized linear mixed model analysis of variance was performed using PROC GLIMMIX. Values are least squares means, and values with different letters are significantly different based on the least squares difference test (α = 0.05).

In 2023, weather conditions were favorable for disease. Tar spot was the most prominent disease in the trial and reached moderate severity. Veltyma sprayed with the CO_2 backpack and the ground rig reduced tar spot severity over the nontreated control on September 6 (Table 25). Veltyma sprayed with the ground rig, the CO_2 backpack, and the drone significantly reduced tar spot severity over the nontreated control, but there was no difference between application type on September 29 and October 16. There was no significant difference in treatments for canopy greenness and yield of corn.

TABLE 25. *Fungicide Application Effect on Tar Spot Severity, Canopy Greenness, and Yield of Corn*

APPLICATION EQUIPMENT AND RATE/ACRE[z]	TAR SPOT[y] % SEPTEMBER 6	TAR SPOT[y] % SEPTEMBER 29	TAR SPOT[y] % OCTOBER 16	CANOPY GREEN[x] %	YIELD[w] BU/ACRE
Nontreated control	1.1 a	2.2 a	8.2 a	38.8	190.4
Ground-rig with Veltyma 3.34 S	0.8 b	1.2 b	3.2 b	52.6	215.7
CO_2 backpack with Veltyma 3.34 S	0.8 b	1.2 b	2.3 b	51.3	217.9
DJI Agras T10 Drone with Veltyma 3.34 S	1.1 a	1.6 b	3.4 b	51.3	214.4
P-value[v]	0.0026	0.0013	0.0028	0.1433	0.1429

[z] Fungicide treatment was applied on August 21 at silk (R1) growth stage using the ground rig, the CO_2 backpack, and the drone. All foliar treatments contained a nonionic surfactant (Preference) at a rate of 0.25% v/v.

[y] Tar spot severity was visually assessed as a percentage (0–100%) of leaf area on five plants in each plot at the ear leaf on September 6, September 29, and October 16 at early dent (R5) and late dent (R5) growth stages.

[x] Canopy greenness was visually assessed as a percentage (0–100%) of canopy green on October 16 at late dent (R5) growth stage.

[w] Yields were adjusted to 15.5% moisture and harvested on November 3.

[v] All data were analyzed in SAS 9.4 (SAS Institute, Cary, NC). A generalized linear mixed model analysis of variance was performed using PROC GLIMMIX. Values are least squares means, and values with different letters are significantly different based on the least squares difference test (α = 0.05).

EVALUATION OF FOLIAR FUNGICIDES IN CORN IN NORTHWESTERN INDIANA, 2023 (COR23-23.PPAC)

N. Yewle, S. Shim, and D. E. P. Telenko, Department of Botany and Plant Pathology, Purdue University West Lafayette, IN 47907-2054

CORN (*ZEA MAYS* W2585VT2PRIB)

Tar spot, *Phyllachora maydis*

A trial was established at the Pinney Purdue Agricultural Center (PPAC) in Porter County, Indiana. The trial was a randomized complete block design with four replications. Plots were 10 feet wide and 30 feet long and consisted of four rows, and the two center rows were used for evaluation. The previous crop was corn. Standard practices for grain corn production in Indiana were followed. Corn hybrid W2585VT2PRIB was planted in 30-inch row spacing at a rate of 34,000 seeds/acre on May 22. The field was overhead irrigated weekly at 1 inch unless weekly rainfall was 1 inch. or higher to encourage disease. All foliar fungicide applications were applied at 15 gal/acre and at 40 psi using a Lee self-propelled sprayer equipped with a 10-foot boom, fitted with six TJ-VS 8002 nozzles spaced 20 inches apart. Fungicides were applied on July 5 and August 2 at V6/V7 and blister (R2) growth stages, respectively. Disease ratings were assessed on September 12 and October 9 at dough (R4) and dent/maturity (R5/R6) growth stages, respectively. Tar spot was rated by visually assessing the percentage (0–100%) of symptomatic leaf area on ear leaf, and five plants were assessed per plot and averaged before analysis. The two center rows of each plot were harvested on November 3, and yields were adjusted to 15.5% moisture. All data were analyzed in SAS 9.4 (SAS Institute, Cary, NC). A generalized linear mixed model analysis of variance was performed using PROC GLIMMIX. Values are least squares means, and values with different letters are significantly different based on the least squares difference test ($\alpha = 0.05$).

In 2023, weather conditions were moderately favorable for tar spot disease. Tar spot was the most prominent disease in the trial and reached moderate severity. Tar spot severity was significantly reduced by Affiance + GWN 10337 at V6/V7, Affiance at R2, Affiance + GWN 10337 at R2, Affiance at V6/V7 followed by Veltyma at R2, Domark at V6/V7 followed by Veltyma at R2, and GWN 10337 + GWN 14510.00001 40.0 fl oz at R2 on September 12 (Table 26). No differences between treatments and the nontreated control were detected for tar spot on October 9. Affiance at V6/V7, Affiance at R2, Affiance + GWN 10337 at R2, Affiance at V6/V7 followed by Veltyma at R2, Domark at V6/V7 followed by Veltyma at R2, and GWN 10337 + GWN 14510.00001 4.0 fl oz at R2 significantly increased canopy greenness over the nontreated control. There was no significant effect of treatment on harvest moisture, test weight, and yield of corn.

TABLE 26. *Effect of Treatments on Tar Spot Severity, Canopy Greenness, and Yield of Corn*

TREATMENT, RATE/ACRE, AND TIMING[z]	TAR SPOT[y] % SEPTEMBER 12	TAR SPOT[y] % OCTOBER 9	CANOPY GREEN[x] %	HARVEST MOISTURE %	TEST WEIGHT LB/BU	YIELD[w] BU/ACRE
Nontreated control	1.3 a	15.6	35.0 de	23.8	52.3	204.2
Affiance 1.5 SC 10.0 fl oz at V6/V7	1.2 ab	17.5	45.0 abc	24.1	51.7	203.5
Affiance 1.5 SC 10.0 fl oz + GWN 10337 2.5 fl oz at V6/V7	0.8 bcd	17.3	43.8 bcd	24.8	51.3	193.1
Affiance 1.5 SC 10.0 fl oz at R2	0.8 bcd	17.1	46.3 abc	23.2	52.4	200.2
Affiance 1.5 SC 10.0 fl oz + GWN 10337 2.5 fl oz at R2	0.5 cd	11.0	50.0 ab	24.9	51.6	205.1
Affiance 1.5 SC 10.0 fl oz at V6/V7 fb Veltyma 3.34 S 7.0 fl oz at R2	0.4 d	12.6	53.8 a	24.0	51.7	215.5
Domark 230 ME 4.0 fl oz at V6/V7 fb Veltyma 3.34 S 7.0 fl oz at R2	0.5 cd	11.7	48.8 ab	24.5	51.5	211.8
Domark 230 ME 4.0 fl oz at + Siapton 24.0 fl oz + Veltyma 3.34 S 7.0 fl oz at V6/V7	0.9 abc	11.6	42.5 bcd	23.7	51.5	208.9
GWN 10337 6.0 fl oz + GWN 14510.00001 4.0 fl oz at R2	0.5 cd	8.1	48.8 ab	24.0	51.6	212.3
GWN 10337 6.0 fl oz + GWN 14510.00001 6.0 fl oz at R2	1.3 ab	14.5	37.5 cde	24.2	53.4	206.5
Affiance 1.5 SC 10.0 fl oz + Badge X2 SC 32.0 fl oz at V6/V7	1.0 ab	19.0	32.5 e	23.5	51.5	200.6
P-value[v]	0.0012	0.3066	0.0018	0.6354	0.8317	0.3810

[z] Foliar fungicides were applied on July 5 and August 2 at V6/V7 and blister (R2) growth stages, respectively. fb = followed by.

[y] Tar spot stromata was visually assessed as a percentage (0–100%) of leaf area on five plants in each plot at the ear leaf on September 12 and October 9 at dough (R4) and dent/maturity (R5/R6) growth stages, respectively.

[x] Canopy greenness was visually assessed as a percentage (0–100%) of crop canopy green on October 9.

[w] Yields were adjusted to 15.5% moisture and harvested on November 3.

[v] All data were analyzed in SAS 9.4 (SAS Institute, Cary, NC). A generalized linear mixed model analysis of variance was performed using PROC GLIMMIX. Values are least squares means, and values with different letters are significantly different based on the least squares difference test (α = 0.05).

EVALUATION OF FUNGICIDES FOR FOLIAR DISEASES IN CORN IN NORTHWESTERN INDIANA, 2023 (COR23-24.PPAC)

E. A. Duncan. S. Shim, and D. E. P. Telenko, Department of Botany and Plant Pathology, Purdue University West Lafayette, IN 47907-2054

CORN (*ZEA MAYS* W2585VT2PRIB)

Tar spot, *Phyllachora maydis*

A trial was established at the Pinney Purdue Agricultural Center (PPAC) in Porter County, Indiana. The experiment was a randomized complete block design with four replications. Plots were 10 feet wide and 30 feet long and consisted of four rows, and the two center rows were used for evaluation. The previous crop was soybean. Standard practices for nonirrigated grain corn production in Indiana were followed. Corn hybrid W2585VT2PRIB was planted in 30-inch row spacing at a rate of 34,000/acre on May 18. At planting, 2x2 applications were made at 10 gal/acre. Foliar applications were made at blister (R2a/R2b), early dough (R4), and late dough (R4) growth stages on August 2, August 3, August 22, and August 29, respectively. All foliar fungicide applications were applied at 15 gal/acre and 40 psi using a Lee self-propelled sprayer equipped with a 10-foot boom, fitted with six TJ-VS 8002 nozzles spaced 20 inches apart. Disease ratings were assessed on September 2, September 14, and October 3 at dough (R4), late dough/early dent (R4/R5), and full dent (R5) growth stages, respectively. Tar spot severity was visually assessed as a percentage (0–100%) of symptomatic leaf area at ear leaf on five plants per plot and averaged before analysis. Percent of canopy greenness was visually assessed as a percentage (0–100%) of canopy green on November 3. The two center rows of each plot were harvested on November 3, and yields were adjusted to 15.5% moisture. All data were analyzed in SAS 9.4 (SAS Institute, Cary, NC). A generalized linear mixed model analysis of variance was performed using PROC GLIMMIX. Values are least squares means, and values with different letters are significantly different based on the least squares difference test (α = 0.05).

In 2023, weather conditions were favorable for disease development. Tar spot was the most prominent disease in the plot and reached moderate severity. Xyway followed by Adastrio at R2a or R2b and Adastrio at R2a, R2b, and early R4 significantly reduced tar spot severity on September 2 over the nontreated control (Table 27). All programs reduced tar spot severity over the nontreated control on September 14. Xyway followed by Adastrio at R2a, R2b, or early R4 and Adastrio at R2a and early R4 reduced tar spot severity as compared to the nontreated control on October 3. Canopy greenness was greatest when Adastrio was applied at early R4. No differences were detected for lodging, moisture, and test weight. Xyway 9.5 fl oz; Xyway followed by Adastrio at R2a, early R4, and late R4; and Adastrio at R2b and early R4 significantly increased yield over the nontreated control (p = 0.10).

TABLE 27. *Effect of Fungicide on Tar Spot Severity, Canopy Greenness, and Yield of Corn*

TREATMENT, RATE/ACRE, AND TIMING[z]	TAR SPOT[y] % SEPTEMBER 2	TAR SPOT[y] % SEPTEMBER 14	TAR SPOT[y] % OCTOBER 3	CANOPY GREEN[x] %	TEST WEIGHT LB/BU	YIELD[w] BU/ACRE
Nontreated control	2.3 a	11.2 a	36.3 a	1.3 c	53.3	190.9 c
Xyway LFR 1.92 SC 9.5 fl oz 2x2	1.5 abc	7.0 b	34.5 ab	4.0 bc	53.4	220.8 a
Xyway LFR 1.92 SC 15.2 fl oz 2x2	1.4 a-d	5.2 b-e	34.6 ab	4.4 bc	53.7	198.2 bc
Xyway LFR 1.92 SC 9.5 fl oz 2x2 fb Adastrio 4.0 SC 9.5 fl oz at R2a	0.5 d	3.6 ef	30.0 cd	4.5 bc	53.5	210.2 ab
Xyway LFR 1.92 SC 9.5 fl oz 2x2 fb Adastrio 4.0 SC 7.0 fl oz at R2b	0.8 bcd	4.5 c-f	32.3 bc	7.3 bc	53.3	196.7 bc
Xyway LFR 1.92 SC 9.5 fl oz 2x2 fb Adastrio 4.0 SC 7.0 fl oz at early R4	1.8 ab	2.7 f	27.0 de	13.0 b	52.6	210.0 ab
Xyway LFR 1.92 SC 9.5 fl oz 2x2 fb Adastrio 4.0 SC 7.0 fl oz at late R4	1.7 ab	6.9 bc	34.5 ab	2.5 bc	53.1	211.7 ab
Adastrio 4.0 SC 8.0 fl oz at R2a	0.6 cd	4.2 def	32.0 bc	9.5 bc	52.7	208.3 abc
Adastrio 4.0 SC 8.0 fl oz at R2b	1.0 bcd	4.7 c-f	34.3 ab	7.8 bc	52.7	219.7 a
Adastrio 4.0 SC 8.0 fl oz at early R4	1.0 bcd	2.8 f	25.3 e	25.0 a	52.7	212.1 ab
Adastrio 4.0 SC 8.0 fl oz at late R4	1.5 abc	5.9 bcd	34.0 ab	2.0 bc	51.7	200.7 bc
P-value[v]	*0.0385*	*0.0001*	*0.0001*	*0.0121*	*0.3145*	*0.0684*

[z] Xyway applications were applied at 2x2 at 10 gal/acre on May 18. Foliar applications were made at blister (R2a/R2b), early dough (R4), and late dough (R4) growth stages on August 2, August 3, August 22, and August 29, respectively. fb = followed by.

[y] Tar spot severity was visually assessed as a percentage (0–100%) of leaf area on five plants in each plot at the ear leaf on September 2, September 14, and October 3 at dough (R4), late dough/early dent (R4/R5), and full dent (R5) growth stages, respectively.

[x] Canopy greenness was visually assessed as a percentage (0–100%) of crop canopy green on October 3.

[w] Yields were adjusted to 15.5% moisture and harvested on November 3.

[v] All data were analyzed in SAS 9.4 (SAS Institute, Cary, NC). A generalized linear mixed model analysis of variance was performed using PROC GLIMMIX. Values are least squares means, and values with different letters are significantly different based on the least squares difference test (α = 0.05).

EVALUATION OF FUNGICIDE PROGRAMS FOR FOLIAR DISEASES IN CORN IN NORTHWESTERN INDIANA, 2023 (COR23-36.PPAC)

N. Yewle, S. Shim, and D. E. P. Telenko, Department of Botany and Plant Pathology, Purdue University West Lafayette, IN 47907-2054

CORN (*ZEA MAYS* W2585VT2PRIB)

Tar spot, *Phyllachora maydis*

A trial was established at the Pinney Purdue Ag Center (PPAC) in Porter County, Indiana. The experiment was a randomized complete block design with four replications. Plots were 10 feet wide and 30 feet long and consisted of four rows, and the two center rows were used for evaluation. The previous crop was soybean. Standard practices for no-till corn production in Indiana were followed. Corn hybrid W2585VT2PRIB was planted in 30-inch row spacing at a rate of 34,000 seeds/acre on May 22. Foliar applications were made at V9/V10 and tassel/silk (VT/R1) growth stages on July 7 and August 3, respectively. All foliar fungicide applications were applied at 15 gal/acre and at 40 psi using a Lee self-propelled sprayer equipped with a 10-foot boom, fitted with six TJ-VS 8002 nozzles spaced 20 inches apart. Disease ratings were assessed on September 14 and October 9 at dent (R5) and physiological maturity (R6) growth stages, respectively. Tar spot severity was visually assessed as a percentage (0–100%) of symptomatic leaf area at ear leaf on five plants per plot and averaged before analysis. Percent of canopy greenness was visually assessed as a percentage (0–100%) of crop canopy green on October 17. The two center rows of each plot were harvested on November 3, and yields were adjusted to 15.5% moisture. All data were analyzed in SAS 9.4 (SAS Institute, Cary, NC). A generalized linear mixed model analysis of variance was performed using PROC GLIMMIX. Values are least squares means, and values with different letters are significantly different based on least squares difference test (α = 0.05).

In 2023, weather conditions were favorable for tar spot disease. Tar spot was the most prominent disease in the trial and reached moderate level. All fungicide programs significantly reduced tar spot severity compared to the nontreated control on September 14 and October 9 (Table 28). All fungicide programs significantly increased canopy greenness over the nontreated control (Table 28). There was no significant effect of treatments for moisture and test weight. Veltyma + OR-009E 0.4 % v/v at VT/R1 and Quadris at V9/V10 followed by Veltyma at VT/R1 and Veltyma + OR-296 at VT/R1 programs significantly increased yield over the nontreated control.

TABLE 28. *Effect of Treatments on Foliar Disease Severity, Canopy Greenness, and Yield of Corn*

TREATMENT, RATE/ACRE, AND TIMING[z]	TAR SPOT %[y] SEPTEMBER 14	TAR SPOT %[y] OCTOBER 9	CANOPY GREEN[x] %	HARVEST MOISTURE %	TEST WEIGHT LB/BU	YIELD[w] BU/ACRE
Nontreated control	3.3 a	26.5 a	30.0 b	24.0	51.9	199.0 c
Veltyma 3.34 S 7.0 fl oz at VT/R1	0.5 b	6.3 b	58.8 a	24.5	52.1	214.3 abc
Veltyma 3.34 S 7.0 fl oz + OR-009E 0.4% v/v at VT/R1	0.3 b	3.0 bc	56.3 a	24.3	52.0	226.3 a
Quadris 2.08 SC 6.0 fl oz at V9/V10 fb Veltyma 3.34 S 7.0 fl oz at VT/R1	0.2 b	3.7 bc	56.3 a	24.4	51.3	224.2 a
Quadris 2.08 SC 6.0 fl oz + OR-484 5.5 fl oz + OR-009E 0.4 % v/v at V9/V10 fb Veltyma 3.34 S 7.0 fl oz + OR-009E 0.40% v/v at VT/R1	0.2 b	1.8 c	61.3 a	25.3	51.5	197.9 cd
Quadris 2.08 SC 6.0 fl oz + OR-009E 0.40% v/v at V9/V10 fb Veltyma 3.34 S 7.0 fl oz + OR-009E 0.40% v/v at VT/R1	0.1 b	2.5 bc	57.5 a	24.5	53.0	200.2 bc
Quadris 2.08 SC 6.0 fl oz + OR-599 +329 0.625% v/v at V9/V10 fb Veltyma 3.34 S 7.0 fl oz + OR-009E 0.40% v/v at VT/R1	0.2 b	2.8 bc	65.0 a	25.5	52.3	181.0 d
Veltyma 3.34 S7.0 fl oz + OR-296 at VT/R1	0.3 b	2.7 bc	60.0 a	25.0	51.8	217.9 ab
P-value[v]	0.0075	0.0001	0.0083	0.0731	0.1743	0.0004

[z] Foliar applications were made at V9/V10 and tassel/silk (VT/R1) growth stages on July 7 and August 3, respectively. All foliar fungicide applications were applied at 15 gal/acre, fb = followed by.

[y] Tar spot stromata was visually assessed as a percentage (0–100%) of leaf area on five plants in each plot at the ear leaf on July 29, September 14, and October 9 at milk (R3), dent (R5), and physiological maturity (R6) growth stages, respectively.

[x] Canopy greenness was visually assessed as a percentage (0–100%) of crop canopy green on October 17.

[w] Yields were adjusted to 15.5% moisture and harvested on November 3.

[v] All data were analyzed in SAS 9.4 (SAS Institute, Cary, NC). A generalized linear mixed model analysis of variance was performed using PROC GLIMMIX. Values are least squares means, and values with different letters are significantly different based on the least squares difference test (α = 0.05).

EVALUATION OF FUNGICIDE PROGRAMS FOR FOLIAR DISEASES IN CORN IN NORTHWESTERN INDIANA, 2023 (COR23-39.PPAC)

C. Rocco da Silva, S. Shim, and D. E. P. Telenko, Department of Botany and Plant Pathology, Purdue University West Lafayette, IN 47907-2054

CORN (*ZEA MAYS* W2585VT2PRIB)

Tar spot, *Phyllachora maydis*

A trial was established at the Pinney Purdue Agricultural Center (PPAC) in Porter County, Indiana. The experiment was a randomized complete block design with four replications. Plots were 10 feet wide and 30 feet long and consisted of four rows, and the two center rows were used for evaluation. The previous crop was corn. Standard practices for nonirrigated grain corn production in Indiana were followed. Corn hybrid W2585VT-2PRIB was planted in 30-inch row spacing at a rate of 34,000 seeds/acre on May 18. In-furrow applications were made at planting at 10 gal/acre. Foliar applications were made at silk (R1) and blister (R2) growth stages on August 2 and August 3, respectively. All foliar fungicide applications were applied at 15 gal/acre and at 40 psi using a Lee self-propelled sprayer equipped with a 10-foot boom, fitted with six TJ-VS 8002 nozzles spaced 20 inches apart. Disease ratings were assessed on August 28, September 13, and October 3 at dough (R4), dent (R5), and physiological maturity (R6), respectively. Tar spot severity was visually assessed as a percentage (0–100%) of symptomatic leaf area at ear leaf on five plants per plot and averaged before analysis. The two center rows of each plot were harvested on November 3, and yields were adjusted to 15.5% moisture. All data were analyzed in SAS 9.4 (SAS Institute, Cary, NC). A generalized linear mixed model analysis of variance was performed using PROC GLIMMIX. Values are least squares means, and values with different letters are significantly different based on the least squares difference test (α = 0.05).

In 2023, weather conditions were favorable for tar spot development. Tar spot was the prominent disease in the trial and reached moderate severity in the trial. Delaro Complete at R1 significantly reduced tar spot on August 28 (Table 29). No significant differences were found between fungicide treatments and the nontreated control for tar spot on September 13 and October 3. There were no significant differences for canopy greenness, harvest moisture, test weight, and yield of corn.

TABLE 29. *Effect of Fungicide on Tar Spot Severity, Canopy Greenness, and Corn Yield*

TREATMENT, RATE/ACRE, AND TIMING[z]	TAR SPOT %[y] AUGUST 28	TAR SPOT %[y] SEPTEMBER 13	TAR SPOT %[y] OCTOBER 3	CANOPY GREEN[x] %	HARVEST MOISTURE %	TEST WEICGHT LB/BU	YIELD[w] BU/ ACRE
Nontreated control	0.7 ab	5.0	10.0	28.8	21.6	53.0	212.5
Tepera Plus HD LFCç 439 L 5.4 fl oz in-furrow	0.8 a	5.0	11.4	20.0	21.4	53.0	206.8
Zolera FX 400 SE 5.0 fl oz at R1	0.7 ab	5.8	10.2	26.3	20.9	53.6	201.5
Zolera FX 400 SE 5.0 fl oz at R2	0.8 a	4.9	12.2	21.3	21.3	53.3	212.9
Tepera Plus HD LFC 439 L 5.4 fl oz in-furrow fb Zolera FX 400 SE 5.0 fl oz at R1	0.6 ab	4.9	10.7	27.5	21.8	52.1	211.9
Tepera Plus HD LFC 439 L 5.4 fl oz in-furrow fb Zolera FX 400 SE 5.0 fl oz at R2	0.4 bc	3.5	9.3	27.5	22.0	52.8	217.0
Delaro Complete 458 SC 8.0 fl oz at R1	0.2 c	2.8	8.3	42.5	22.5	52.6	220.0
P-value[v]	*0.0189*	*0.2765*	*0.5870*	*0.4290*	*0.6707*	*0.7967*	*0.3747*

[z] In-furrow applications was made at planting at 10 gal/acre. Foliar applications were made at silk (R1) and blister (R2) growth stages on August 2 and August 3, respectively. fb = followed by.

[y] Tar spot stromata was visually assessed as a percentage (0–100%) of leaf area on five plants in each plot at the ear leaf on August 28, September 13, and October 3 at dough (R4), dent (R5), and physiological maturity (R6), respectively.

[x] Canopy greenness was visually assessed as a percentage (0–100%) of crop canopy on October 3.

[w] Yields were adjusted to 15.5% moisture and harvested on November 3.

[v] All data were analyzed in SAS 9.4 (SAS Institute, Cary, NC). A generalized linear mixed model analysis of variance was performed using PROC GLIMMIX. Values are least squares means, and values with different letters are significantly different based on the least squares difference test (α = 0.05).

EVALUATION OF FUNGICIDES FOR FOLIAR DISEASES IN CORN IN NORTHWESTERN INDIANA, 2023 (COR23-41.PPAC)

E. A. Duncan, S. Shim, and D. E. P. Telenko, Department of Botany and Plant Pathology, Purdue University West Lafayette, IN 47907-2054

CORN (*ZEA MAYS* W2585VT2PRIB)

Tar spot, *Phyllachora maydis*

A trial was established at the Pinney Purdue Agricultural Center (PPAC) in Porter County, Indiana. The experiment was a randomized complete block design with four replications. Plots were 10 feet wide and 30 feet long and consisted of four rows, and the two center rows were used for evaluation. The previous crop was corn. Standard practices for grain corn production in Indiana were followed. Corn hybrid W2385VT2PRIB was planted in 30-inch row spacing at a rate of 2 seeds/foot on May 18. The field was overhead irrigated weekly at 1 inch unless weekly rainfall was 1inch or more to encourage disease. Foliar applications were made at V10, blister (R2), and dough (R4) growth stages on July 17, August 2, and August 22, respectively. All foliar fungicide applications were applied at 15 gal/acre and 40 psi using a Lee self-propelled sprayer equipped with a 10-foot boom, fitted with six TJ-VS 8002 nozzles spaced 20 inches apart. Disease ratings were assessed on August 30, September 14, and October 9 at dough (R4), early dent (R5), and late dent (R5) growth stages, respectively. Tar spot severity was visually assessed as a percentage (0–100%) of symptomatic leaf area at ear leaf on five plants per plot and averaged before analysis. Percent of canopy greenness was visually assessed as a percentage (0–100%) of crop canopy green on October 9. The two center rows of each plot were harvested on November 3, and yields were adjusted to 15.5% moisture. All data were analyzed in SAS 9.4 (SAS Institute, Cary, NC). A generalized linear mixed model analysis of variance was performed using PROC GLIMMIX. Values are least squares means, and values with different letters are significantly different based on the least squares difference test ($\alpha = 0.05$).

In 2023, weather conditions were favorable for disease development. Tar spot was the most prominent disease and reached moderate severity. No significant differences were detected for tar spot severity on August 30 (Table 30). Xyway 9.5 fl oz 2x2 followed by Veltyma at R4 and Veltyma at R2 significantly reduced tar spot over the nontreated control on September 14. All fungicide programs significantly reduced tar spot over the nontreated control on October 9. Veltyma at R2 had significantly increased canopy greenness compared to the nontreated control on October 9. No differences between fungicide programs and the nontreated control were detected for test weight and yield of corn.

TABLE 30. *Effect of Fungicide Treatment on Tar Spot Severity, Canopy Greenness, and Yield of Corn*

TREATMENT, RATE/ACRE, AND TIMING[z]	TAR SPOT %[y] AUGUST 30	TAR SPOT %[y] SEPTEMBER 14	TAR SPOT %[y] OCTOBER 9	CANOPY GREEN[x] %	TEST WEIGHT LB/BU	YIELD[w] BU/ACRE
Nontreated control	0.15	0.8 ab	19.3 a	36.3 bc	60.7	201.9
Xyway LFR 1.92 SC 9.5 fl oz in 2x2	0.11	0.8 ab	7.9 b	33.3 bc	51.5	204.9
Xyway LFR 1.92 SC 15.2 fl oz in 2x2	0.11	1.0 a	8.1 b	28.8 c	52.1	206.6
Xyway LFR 1.92 SC 9.5 fl oz in 2x2 fb Adastrio 4.0 SC 7.0 fl oz at R2	0.06	0.7 abc	8.9 b	37.5 bc	52.0	213.8
Xyway LFR 1.92 SC 9.5 fl oz in 2x2 fb Adastrio 4.0 SC 7.0 fl oz at R4	0.05	0.5 bc	6.9 b	33.8 bc	51.9	219.7
Xyway LFR 1.92 SC 9.5 fl oz in 2x2 fb Veltyma 3.34 S 7.0 fl oz at R4	0.13	0.4 c	4.3 b	41.7 ab	50.9	210.5
Topguard 1.04 SC 10.0 fl oz at V10 fb Adastrio 4.0 SC 8.0 fl oz at R4	0.09	0.6 bc	7.7 b	33.4 bc	52.5	214.7
Adastrio 8.0 fl oz at R2	0.09	0.6 abc	7.4 b	40.0 bc	50.4	208.9
Delaro 325 SC 5.0 fl oz at V10 fb Delaro Complete 458 SC 8.0 fl oz at R4	0.06	0.6 bc	3.9 b	45.0 ab	51.7	206.5
Veltyma 3.34 S 7.0 fl oz at R2	0.04	0.3 c	4.7 b	52.5 a	60.6	218.7
P-value[v]	*0.0975*	*0.0331*	*0.0009*	*0.0147*	*0.5688*	*0.4305*

[z] Foliar applications were made at V10, blister (R2), and dough (R4) growth stages on July 17, August 2, and August 22, respectively. All foliar fungicide treatments at blister (R2) and dough (R4) contained a nonionic surfactant (Preference) at a rate of 0.25% v/v. fb = followed by.

[y] Tar spot stromata was visually assessed as a percentage (0–100%) of leaf area on five plants in each plot at the ear leaf on August 30, September 14, and October 9 at dough (R4), early dent (R5), and late dent (R5) growth stages, respectively.

[x] Canopy greenness was visually assessed as a percentage (0–100%) of crop canopy green on October 9.

[w] Yields were adjusted to 15.5% moisture and harvested on November 3.

[v] All data were analyzed in SAS 9.4 (SAS Institute, Cary, NC). A generalized linear mixed model analysis of variance was performed using PROC GLIMMIX. Values are least squares means, and values with different letters are significantly different based on the least squares difference test (α = 0.05).

EVALUATION OF FUNGICIDES FOR FOLIAR DISEASES IN CORN IN NORTHWESTERN INDIANA, 2023 (COR23-44.PPAC)

E. A. Duncan, S. Shim, and D. E. P. Telenko, Department of Botany and Plant Pathology, Purdue University West Lafayette, IN 47907-2054

CORN (*ZEA MAYS* W2585VT2PRIB)

Tar spot, *Phyllachora maydis*

A trial was established at the Pinney Purdue Agricultural Center (PPAC) in Porter County, IN. The trial was a randomized complete block design with four replications. Plots were 10 feet wide and 30 feet long and consisted of four rows, and the two center rows were used for evaluation. The previous crop was corn. Standard practices for grain corn production in Indiana were followed. Corn hybrid W2585VT2PRIB was planted in 30-inch row spacing at a rate of 34,000 seeds/acre on May 25. Foliar applications were made at V12, blister (R2), and dough (R4) growth stages on July 25, August 2, and August 22, respectively. All foliar fungicide applications were applied at 15 gal/acre and 40 psi using a Lee self-propelled sprayer equipped with a 10-foot boom, fitted with six TJ-VS 8002 nozzles spaced 20 inches apart. Disease ratings were assessed in August 28, September, and October at dough (R4), dent (R5), and physiological maturity (R6), respectively. Tar spot severity was visually assessed as a percentage (0–100%) of symptomatic leaf area at ear leaf on five plants per plot and averaged before analysis. Percent of canopy greenness was visually assessed as a percentage (0–100%) of crop canopy green on October 3. The two center rows of each plot were harvested on November 6, and yields were adjusted to 15.5% moisture. All data were analyzed in SAS 9.4 (SAS Institute, Cary, NC). A generalized linear mixed model analysis of variance was performed using PROC GLIMMIX. Values are least squares means, and values with different letters are significantly different based on the least squares difference test ($\alpha = 0.05$).

In 2023, weather conditions were favorable for disease development. Tar spot was the most prominent disease in the trial and reached moderate severity. All fungicides significantly reduced tar spot at ear leaf on August 28 over the nontreated control except Miravis Neo at R4 (Table 31). All fungicide applications reduced tar spot on September 12 over the nontreated control. Miravis Neo applied at R2 followed by Miravis Neo at R4 and Miravis Neo at R2 followed by Quilt Xcel at R4 resulted in the lowest level of tar spot on September 12. On October 3 all fungicide programs reduced tar spot stromata over the nontreated control. Canopy greenness was significantly higher than the nontreated control with Veltyma at V12 and R2, Delaro Complete at R2, and Miravis Neo at R2 followed by either Miravis Neo or Quilt Xcel at R4 (Table 31). No significant differences were detected for test weight and yield of corn.

TABLE 31. *Effect of Fungicide Treatment on Tar Spot Severity, Canopy Greenness, and Yield of Corn*

TREATMENT, RATE/ACRE, AND TIMING[z]	TAR SPOT %[y] AUGUST 28	TAR SPOT %[y] SEPTEMBER 12	TAR SPOT %[y] OCTOBER 3	CANOPY GREEN[x] %	TEST WEIGHT LB/BU	YIELD[w] BU/ACRE
Nontreated control	0.23 a	0.80 a	3.9 a	56.3 b	54.8	234.9
Miravis Neo 2.5 EC 13.7 fl oz at V12	0.05 bc	0.51 bc	2.3 bc	65.7 ab	51.8	231.6
Veltyma 3.34 S 7.0 fl oz at V12	0.00 c	0.39 bc	1.5 bcd	80.0 a	51.7	243.6
Miravis Neo 2.5 EC 13.7 fl oz at R2	0.04 c	0.53 b	2.5 b	71.3 ab	52.6	240.8
Adastrio 4.0 SC 7.0 fl oz at R2	0.02 c	0.43 bc	2.0 bc	68.8 ab	52.8	247.8
Veltyma 3.34 S 7.0 fl oz at R2	0.01 c	0.37 bc	1.4 bcd	75.0 a	51.4	238.6
Delaro Complete 458 SC 8.0 fl oz at R2	0.00 c	0.43 bc	1.6 bcd	80.0 a	52.0	252.1
Miravis Neo 2.5 EC 13.7 fl oz at R4	0.14 ab	0.26 cd	0.9 cde	65.0 ab	52.0	240.7
Miravis Neo 2.5 EC 13.7 fl oz at R2 fb Miravis Neo 2.5 EC 13.7 fl oz at R4	0.00 c	0.04 d	0.3 e	80.0 a	52.7	239.4
Miravis Neo 2.5 EC 13.7 fl oz at R2 fb Quilt Xcel 2.2 SE 10.5 fl oz at R4	0.00 c	0.09 d	0.6 de	75.0 a	52.1	244.9
P-value[v]	*0.0001*	*0.0002*	*0.0001*	*0.0482*	*0.4949*	*0.2591*

[z] Foliar fungicide applications were made at V12, blister (R2), and dough (R4) growth stages on July 25, August 2, and August 22, respectively. All foliar fungicide applications were applied at 15 gal/acre, and applications made after VT/R1 contained a nonionic surfactant (Preference) at a rate of 0.25% v/v. fb = followed by.

[y] Tar spot stromata was visually assessed as a percentage (0–100%) of leaf area on five plants in each plot at the ear leaf on August 28, September 12, and October 3 at dough (R4), dent (R5), and physiological maturity (R6), respectively.

[x] Canopy greenness was visually assessed as a percentage (0–100%) of crop canopy green on October 3.

[w] Yields were adjusted to 15.5% moisture and harvested on November 6.

[v] All data were analyzed in SAS 9.4 (SAS Institute, Cary, NC). A generalized linear mixed model analysis of variance was performed using PROC GLIMMIX. Values are least squares means, and values with different letters are significantly different based on the least squares difference test (α = 0.05).

EVALUATION OF PLANTING DATES AND SEED TREATMENTS ON SOYBEAN IN NORTHWESTERN INDIANA, 2023 (SOY23-11.PPAC)

I. L. Miranda, S. Shim, and D. E. P. Telenko, Department of Botany and Plant Pathology, Purdue University West Lafayette, IN 47907-2054

SOYBEAN (*GLYCINE MAX*)

Septoria brown spot, *Septoria glycines*

A trial was established at Pinney Purdue Agricultural Center (PPAC) in Porter County, Indiana. The experiment was a randomized complete block design with four replications. Plots were 10 feet wide and 40 feet long and consisted of four rows, and the two center rows were used for evaluation. The previous crop was corn. Standard practices for soybean production in Indiana were followed. Soybean seeds were planted in 30-inch row spacing at a rate of 140,000 seeds/acre. Treatments were a factorial arrangement of four planting dates by four seed treatments. Soybeans were planted on April 13, April 27, May 11, and May 31. Stand counts were assessed at cotyledons expanded/first-node stage (VC/V1) growth stage for each planting date. Disease ratings were assessed on September 6 at full seed/beginning maturity (R6/R7) growth stage. Septoria brown spot (SBS) was rated for disease severity by visually assessing the percentage (0–100%) of canopy disease symptoms in each plot. Ten roots were sampled from the outer rows of each plot and rated for root rot severity on a scale of 0–100% and averaged before analysis. Root dry weight was calculated from the 10 root samples. The two center rows of each plot were harvested on October 9, and yields were adjusted to 13% moisture. All data were analyzed in SAS 9.4 (SAS Institute, Cary, NC). A generalized linear mixed model analysis of variance was performed using PROC GLIMMIX. Values are least squares means, and values with different letters are significantly different based on a least squares difference test (α = 0.05).

In 2023, weather conditions were not favorable for disease. SBS was the most prominent disease and reached low severity. There was a significant interaction between planting date and seed treatment for stand count, but no interactions were detected for other variables; therefore, only main effects are presented. Soybean stand counts were the highest in the May 31 planting date with the nontreated control and CruiserMaxx APX without thiamethoxam seed treatments, whereas the planting dates on April 13 and April 27 resulted in the lowest stand across all seed treatments (data not shown). Planting on April 13 resulted on the highest incidence of SBS as compared to later planting dates (Table 32). There were no significant differences between planting dates for root rot severity. Root dry weight was highest at planting dates on April 13 and April 27 as compared to May 11 and May 31. CruiserMaxx APX with thiamethoxam seed treatment also significantly increased root weight compared to all the other seed treatments and the nontreated control. Test weight was significantly higher at planting on May 31 compared to other planting dates. Soybean yield was significantly reduced on the last planting date, May 31, as compared to earlier planting dates. No significant differences were detected between seed treatments for SBS, root rot, test weight, and yield of soybean.

TABLE 32. *Effect of Planting Dates and Seed Treatments on Stand Count, SBS Severity, Root Rot, Root Weight, and Soybean Yield*

PLANTING DATES AND SEED TREATMENT[z]	STAND COUNT #ACRE	SBS %[y]	ROOT ROT[x] %	ROOT DRY WEIGHT[w] G	TEST WEIGHT LB/BU	YIELD[v] BU/ACRE
Planting date (April 13)	85,105	17.5 a	0.5	31.5 a	55.5 b	73.6 a
Planting date (April 27)	89,135	14.1 b	0.3	33.2 a	55.5 b	72.8 a
Planting date (May 11)	138,848	9.4 c	0.2	26.6 b	55.6 b	71.0 a
Planting date (May 31)	152,514	5.3 d	0.3	26.9 b	56.1 a	65.3 b
Nontreated control	119,899	11.6	0.3	28.2 b	55.6	70.8
CruiserMaxx APX + thiamethoxam	111,514	11.3	0.5	33.1 a	55.8	70.6
Thiamethoxam	114,345	11.3	0.2	29.9 b	55.6	71.1
CruiserMaxx APX no thiamethoxam	119,844	12.2	0.4	27.0 b	55.6	70.0
P-value *planting date*[u]	0.0001	0.0001	0.4654	0.0001	0.0023	0.0001
P-value *seed treatment*	0.0413	0.7880	0.4267	0.0019	0.7905	0.8905
P-value *planting date*seed treatment*	0.0060	0.1756	0.6512	0.1594	0.8616	0.8611

[z] Seed treatments applied prior to planting at 10 g AI/100 kg seed.

[y] Disease severity was visually assessed as a percentage (0–100%) of canopy within a plot with symptoms on September 6 at full seed/beginning maturity (R6/R7) growth stage. SBS = Septoria brown spot.

[x] Ten roots per plot were sampled from border rows and gently washed, and root rot was visually assessed as a percentage (0–100%) of dark discoloration on roots on September 8.

[w] Root dry weight was assessed as an average of weight of 10 dried root samples.

[v] Yields were adjusted to 13% moisture and harvested on October 9.

[u] All data were analyzed in SAS 9.4 (SAS Institute, Cary, NC). A generalized linear mixed model analysis of variance was performed using PROC GLIMMIX. Values are least squares means, and values with different letters are significantly different based on the least squares difference test (α = 0.05).

EVALUATION OF FUNGICIDES FOR WHITE MOLD IN SOYBEAN IN NORTHWESTERN INDIANA, 2023 (SOY23-17.PPAC)

E. Duncan, S. Shim, and D. E. P. Telenko, Department of Botany and Plant Pathology, Purdue University
West Lafayette, IN 47907-2054

SOYBEAN (*GLYCINE MAX* P29A19E)

White mold, *Sclerotinia sclerotiorum*

A trial was established at the Pinney Purdue Agricultural Center (PPAC) in Porter County, Indiana. The experiment was a randomized complete block design with four replications. Plots were 10 feet wide and 30 feet long and consisted of four rows, and the two center rows were used for evaluation. The previous crop was corn. Standard practices for soybean production in Indiana were followed. Soybean cultivar P29A19E was planted in 30-inch row spacing at a rate of 140,000 seeds/acre on May 22. Inoculum of *Sclerotinia sclerotiorum* was applied on the seedbed at 1.25 g/foot at planting. The field was overhead irrigated weekly at 1 inch unless weekly rainfall was 1 inch or higher to encourage disease. All fungicide applications were applied at 15 gal/acre and 40 psi using a Lee self-propelled sprayer equipped with a 10-foot boom, fitted with six TJ-VS 8002 nozzles spaced 20 inches apart. Fungicides were applied on July 17 and July 25 at beginning bloom (R1) and beginning pod (R3) growth stages, respectively. White mold disease incidence was assessed by counting the number of plants in each plot with symptoms. For disease severity, each plant that was observed was rated according to the following disease category: 0 = no disease, 1 = lateral branches with white mycelium and lesions, 2 = main stem with white mycelium and sclerotia present, and 3 = entire plant wilted/plant death. The disease severity index (DSI) was calculated by multiplying the average number of plants in each severity category by the incidence: DSI = [sum (disease severity score × number of plants)]/[(maximum disease score) × (disease incidence)] × 100. The center rows of each plot were harvested on October 9, and yields were adjusted to 13% moisture. All data were analyzed in SAS 9.4 (SAS Institute, Cary, NC). A generalized linear mixed model analysis of variance was performed using PROC GLIMMIX. Values are least squares means, and values with different letters are significantly different based on the least squares difference test (α = 0.05).

In 2023, weather conditions were moderately favorable for disease. White mold was the most prominent disease and reached moderate severity. There were no significant differences between fungicide treatments and nontreated control for disease ratings on September 14 (Table 33). Delaro Complete applied at R1 followed by R3 significantly reduced white mold DSI when compared only with nontreated control but was not significantly different for all other treatments. There was no significant effect of treatment on canopy greenness and soybean yield.

TABLE 33. *Effect of Fungicide on White Mold, Canopy Greenness, and Yield of Soybean*

TREATMENT AND RATE/ACRE[z]	WHITE MOLD[y] DIX %	WHITE MOLD[y] DSI %	CANOPY GREEN[x] %	HARVEST MOISTURE %	TEST WEIGHT LB/BU	YIELD[w] BU/ACRE
Nontreated control	34.3	21.6	58.8	12.5	57.2	51.3
Delaro Complete 458 SC 8.0 fl oz at R1	24.3	11.4	68.8	12.5	57.2	53.2
Delaro Complete 458 SC 8.0 fl oz at R1 fb Delaro Complete 458 SC 8.0 fl oz at R3	18.3	9.1	67.5	12.7	56.9	53.3
Miravis Neo 2.5 SE 13.7 fl oz at R1	23.8	11.8	65.0	12.7	56.8	53.5
Endura 70 WDG 6.0 fl oz at R1	28.0	15.4	55.0	12.9	57.3	52.8
Revytek 3.33 LC 8.0 fl oz at R1	25.3	13.1	65.0	12.8	56.9	52.1
Omega 500 F 16.0 fl oz at R1	27.8	14.9	61.3	12.6	57.2	52.9
P-value[v]	0.7638	0.5512	0.7076	0.6628	0.5732	0.8990
P-value *NTC vs. R1 fb R3 program*	0.1038	0.0388	0.3349	0.3222	0.2160	0.2929

[z] Fungicides were applied on July 17 and July 25 at beginning bloom (R1) and beginning pod (R3) growth stages, respectively. All plots were inoculated with *S. sclerotiorum*. fb = followed by.

[y] The disease severity index (DIX) was calculated by multiplying the average number of plants in each severity category by the incidence: DSI = [sum (disease severity score × number of plants)]/[(maximum disease score) × (disease incidence)] × 100.

[x] Canopy greenness was visually assessed as a percent (0–100%) on September 14.

[w] Yields were adjusted to 13% moisture and harvested on October 9.

[v] All data were analyzed in SAS 9.4 (SAS Institute, Cary, NC). A generalized linear mixed model analysis of variance was performed using PROC GLIMMIX. Values are least squares means, and values with different letters are significantly different based on the least squares difference test (α = 0.05).

UNIFORM FUNGICIDE TRIALS FOR WHITE MOLD IN SOYBEAN IN NORTHWESTERN INDIANA, 2023 (SOY23-25.PPAC)

N. Yewle, S. Shim, and D. E. P. Telenko, Department of Botany and Plant Pathology, Purdue University West Lafayette, IN 47907-2054

SOYBEAN (*GLYCINE MAX* P29A19E)

White mold, *Sclerotinia sclerotiorum*

A trial was established at the Pinney Purdue Agricultural Center (PPAC) in Porter County, Indiana. The experiment was a randomized complete block design with four replications. Plots were 10 feet wide and 30 feet long and consisted of four rows, and the two center rows were used for evaluation. The previous crop was corn. Standard practices for soybean production in Indiana were followed. Soybean cultivar P29A19E was planted in 30-inch row spacing at a rate of 140,000 seeds/acre on May 22. Inoculum of *Sclerotinia sclerotiorum* was applied on the seedbed at 1.25 g/foot at planting. The field was overhead irrigated weekly at 1 inch unless weekly rainfall was 1 inch or higher to encourage disease. Treatments were applied using a Lee self-propelled sprayer with a 10-foot boom, using six TJ-VS 8002 nozzles spaced 20 inches apart, applying at 15 gal/acre at 40 psi. In addition, 360 undercover treatments were applied using a CO_2 backpack sprayer with a 10-foot boom, using four 360 nozzles spaced 30 inches apart in 15 gal/acre at 40 psi. Fungicides were applied on July 6, July 17, and July 25 at the V4, beginning bloom (R1), and beginning pod (R3) growth stages, respectively. Sporecaster applications occurred on July 19 at full bloom (R2) growth stage. Disease rating was assessed on September 15 at full seed (R6) growth stage. White mold disease incidence was assessed by counting the number of plants in each plot with symptoms. For disease severity, each plant observed was rated according to the following disease category: 0 = no disease, 1 = lateral branches with white mycelium and lesions; 2 = main stem with white mycelium and sclerotia present, and 3 = entire plant wilted/plant death. The disease severity index (DIX) was calculated by multiplying the average number of plants in each severity category by the incidence: DIX = [sum (disease severity score × number of plants)] / [(maximum disease score) × (disease incidence)] × 100. Canopy greenness was visually assessed as a percentage (0–100%) of canopy green on September 15. The two center rows of each plot were harvested on October 9, and yields were adjusted to 13% moisture. All disease and yield data were analyzed using a generalized linear mixed model analysis of variance that was performed using PROC GLIMMIX. Values are least squares means, and values with different letters are significantly different based on the least squares difference test ($\alpha = 0.05$).

In 2023, weather conditions were favorable for disease. White mold was present in the trial and reached a moderate level. There were no significant differences between fungicide treatments and nontreated control for white mold incidence and index (Table 34). No significant differences were detected between treatments and nontreated control for canopy greenness and yield of soybean.

TABLE 34. *Effect of Fungicide on White Mold Incidence, Canopy Greenness, and Yield of Soybean*

TREATMENT, RATE/ACRE, AND TIMING[z]	WHITE MOLD[y] DI %	WHITE MOLD[y] DIX %	CANOPY GREEN[x] %	YIELD[w] BU/ACRE
Nontreated control	37.5	12.5	84.5	59.6
Endura 70 WDG 8.0 oz at R1 and R3	5.0	1.7	88.5	62.0
Endura 70 WDG 8.0 oz at R3	21.0	7.0	82.0	62.6
Omega 500 F 16.0 fl oz at R3 by 360 under cover	28.5	9.5	73.8	60.2
Omega 500 F 16.0 fl oz at R3	24.3	8.1	90.0	64.1
Cobra 2 EC 8.0 fl oz at V4	27.8	9.3	80.0	59.0
Cobra 2 EC 8.0 fl oz at V4 fb Domark 230 ME 5.0 fl oz at R3	25.3	8.4	86.3	60.5
Omega 500 F 12.0 fl oz at R1 fb Miravis Neo 2.5 SE 13.7 fl oz at R3	10.5	3.5	87.5	62.2
Delaro Complete 458 SC 8.0 fl oz at R3	12.0	4.0	83.3	62.6
Delaro Complete 458 SC 8.0 fl oz at R3 by 360 under cover	26.8	8.9	80.0	58.0
Headsup Seed Treatment	21.8	7.3	80.0	60.6
Headsup Seed Treatment fb Domark 203 ME 5.0 fl oz at R3	43.5	14.5	82.5	58.9
Miravis Neo 2.5 SE 16.0 fl oz at R3	46.0	15.3	90.8	61.0
Phostrol 6.26 SL 4.0 pt + Topsin 4.5 FL 20.0 fl oz at R3	25.0	8.3	78.8	59.7
Omega 500 F 16.0 fl oz with Sporecaster 360 under cover at R2	11.8	3.9	81.3	63.4
Endura 70 WDG 8.0 oz at Sporecaster at R2	15.5	5.2	83.8	61.7
P-value[v]	*0.0621*	*0.0622*	*0.4691*	*0.2559*

[z] Fungicides were applied on July 6, July 17, July 19, and July 25 at the V4, beginning bloom (R1), full bloom (R2) at Sporecaster and beginning pod (R3) growth stages, respectively. All fungicide treatments contained a nonionic surfactant (Preference) at a rate of 0.25% v/v except Cobra. All plots were inoculated with *S. sclerotiorum* at 1.25 g/foot within the seedbed at planting. fb = followed by.

[y] White mold disease incidences were assessed by counting the number of plants/plots with symptoms on September 15. Disease severity index (DIX) is calculated by multiplying the average number of plants in each severity category by the incidence. DIX = [sum (disease severity score × number of plants)]/[(maximum disease score) × (disease incidence)] × 100.

[x] Canopy greenness was visually assessed as a percentage (0–100%) of canopy green on September 15.

[w] Yields were adjusted to 13% moisture and harvested on October 9.

[v] All disease and yield data were analyzed using a generalized linear mixed model analysis of variance performed using PROC GLIMMIX. Values are least squares means, and values with different letters are significantly different based on the least squares difference test (α = 0.05).

EVALUATION OF DISEASE MANAGEMENT PROGRAMS FOR WHITE MOLD IN ORGANIC SOYBEAN IN NORTHWESTERN INDIANA, 2023 (SOY23-26.PPAC)

C. Rocco da Silva, S. Shim, and D. E. P. Telenko, Department of Botany and Plant Pathology, Purdue University West Lafayette, IN 47907-2054

SOYBEAN (*GLYCINE MAX* DWIGHT AND MN1410)

White mold, *Sclerotinia sclerotiorum*

A trial was established at the Pinney Purdue Agricultural Center (PPAC) in Porter County, Indiana. The experiment was a split-plot design with four replications. Main plots were cover crop termination (full tillage vs. roller-crimped rye). Subplots were cultivar by fungicide program and were 6.7 feet wide and 30 feet long and consisted of four rows, and the two center rows were used for evaluation. The previous crop was sunflower. Cereal rye was planted on September 17, 2022, at a rate of 150 lbs/acre. On May 25 the cover crop was terminated using either tillage or roller-crimping. Standard practices for soybean organic production in Indiana were followed. Organic soybean cultivars Dwight and MN1410 were planted in 20-inch row spacing at a rate of 8 seeds/foot on May 25. Inoculum of *Sclerotinia sclerotiorum* was applied within the seedbed at 1.25 g/foot at planting, and 60 sclerotia per plot were spread between the middle two rows after tillage and before roller-crimping. The field was overhead irrigated weekly at 1 inch unless weekly rainfall was 1 inch or higher to encourage disease. All fungicide applications were applied at 15 gal/acre and 40 psi using a Lee self-propelled sprayer equipped with a 10-foot boom, fitted with six TJ-VS 8002 nozzles spaced 20 inches apart. Fungicides were applied on July 25 at full bloom (R2) growth stage. Disease ratings were assessed on September 18 at full seed (R6) growth stage. White mold disease incidence was assessed by counting the number of plants in each plot with symptoms. For disease severity, each plant that is observed should be rated according to the following disease category: 0 = no disease, 1 = lateral branches with white mycelium and lesions, 2 = main stem with white mycelium and sclerotia present, and 3 = entire plant wilted/plant death. The disease severity index (DIX) was calculated by multiplying the average number of plants in each severity category by the incidence: DIX = [sum (disease severity score ×number of plants)]/[(maximum disease score) × (disease incidence)] × 100. The center rows of each plot were harvested on October 10, and yields were adjusted to 13% moisture. All disease and yield data were analyzed in SAS 9.4 (SAS Institute, Cary, NC). A generalized linear mixed model analysis of variance was performed using PROC GLIMMIX. Values are least squares means, and values with different letters are significantly different based on the least squares difference test (α = 0.05).

In 2023, low disease developed in plots. White mold was the most prominent disease in the trial but only reached low severity. The main effects of cultivar, cover crop termination, and fungicide treatments are presented where there were no significant interactions and simple effects where an interaction was detected (Table 35). Full tillage reduced white mold DIX when compared to roller-crimped rye. White mold was lowest in the nontreated Dwight and MN11410 treated with Actinovate, but these were not significant from the MN1410 nontreated. The percentage of green canopy was highest when treated with Endura in roller-crimped rye, but this was not significantly different from full till + Actinovate. There was no significant difference in treatments for white mold disease incidence, defoliation, and soybean yield.

TABLE 35. *Effect of Fungicide on White Mold Incidence, Index, Canopy Greenness, Defoliation, and Yield of Soybean*

TREATMENT[z]	WHITE MOLD DI %[y]	WHITE MOLD DIX[x]		CANOPY GREEN %[w]		DEFOLIATION[v] %	YIELD[u] BU/ ACRE
Cover crop termination							
Full tillage	2.2	2.0 b				68.6	52.0
Roller-crimped rye	3.5	3.3 a				76.8	52.4
Cultivar							
Dwight	3.0			9.0		76.7	52.3
MN1410	2.7			13.0		68.8	52.1
Fungicide programs and rate/acre		Dwight	MN1410	Full till	RCR		
Nontreated control	2.4	0.7 c	3.9 abc	5.4 c	5.2 c	76.7	51.3
Endura 70 WDG 8.0 fl oz	2.2	2.5 abc	1.5 bc	12.6 bc	34.3 a	57.2	50.6
Double Nickel 55 DWG 2 qt	3.6	6.0 a	1.0 bc	5.4 c	2.6 c	85.4	54.1
Serifel WP 16 fl oz	3.7	1.8 bc	4.7 ab	22.5 ab	4.9 c	62.9	51.2
Actinovate AG 12 oz	2.6	4.0 abc	0.6 c	4.1 c	16.6 bc	82.7	53.6
BotryStop 2 lb	2.7	2.0 bc	2.9 abc	14.4 bc	4.1 c	71.4	52.4
P-value till[t]	*0.0542*	**0.0397**		*0.7881*		*0.3279*	*0.5269*
P-value cultivar	*0.6979*	*0.6358*		*0.2476*		*0.2137*	*0.8635*
P-value fungicide	*0.8795*	*0.8656*		*0.0242*		*0.0816*	*0.1829*
P-value till*cultivar	*0.5781*	*0.6146*		*0.9575*		*0.6461*	*0.2624*
P-value till*fungicide	*0.5396*	*0.5360*		**0.0182**		*0.1560*	*0.2713*
P-value cultivar*fungicide	*0.0259*	**0.0193**		*0.6173*		*0.8520*	*0.9800*
P-value till*cultivar*fungicide	*0.5375*	*0.4413*		*0.4274*		*0.4213*	*0.7638*

[z] Fungicide applications were made on July 25 at full bloom (R2) growth stage. All plots were inoculated with *S. sclerotiorum* at 1.25 g/foot within the seedbed at planting and 60 sclerotia per plot were spread between the middle two rows before roller-crimped and after tillage.

[y] White mold disease incidence (DI) % was assessed by counting the number of plants in each plot with symptoms.

[x] The disease severity index (DIX) is calculated by multiplying the average number of plants in each severity category by the incidence: DIX = [sum (disease severity score X number of plants)]/[(maximum disease score) ×(disease incidence)] × 100.

[w] Canopy greenness was visually assessed as a percentage (0–100%) of green of the two center rows on September 29.

[v] Defoliation was assessed as a percentage of leaf loss in plot.

[u] Yields were adjusted to 13% moisture and harvested on October 10.

[t] All disease and yield data were analyzed in SAS 9.4 (SAS Institute, Cary, NC). A generalized linear mixed model analysis of variance was performed using PROC GLIMMIX. Values are least squares means, and values with different letters are significantly different based on the least squares difference test (α = 0.05).

SOUTHWEST PURDUE AGRICULTURAL CENTER (SWPAC)

FUNGICIDE COMPARISON FOR FOLIAR DISEASES IN CORN IN SOUTHERN INDIANA, 2023 (COR23-11.SWPAC)

N. Yewle, S. Shim, and D. E. P. Telenko, Department of Botany and Plant Pathology, Purdue University West Lafayette, IN 47907-2054

CORN (*ZEA MAYS* P0574AM)

Gray leaf spot, *Cercospora zeae-maydis*

A trial was established at the Southwest Purdue Agricultural Center (SWPAC) in Knox County, Indiana. The trial was a randomized complete block design with four replications. Plots were 10 feet wide and 30 feet long and consisted of four rows, and the two center rows were used for evaluation. The previous crop was soybean. Standard practices for grain corn production in Indiana were followed. Corn hybrid P0574AM was planted in 30-inch row spacing at a rate of 27,000 seeds/acre on April 11. All foliar fungicide applications were applied at 15 gal/acre and 40 psi using a Lee self-propelled sprayer equipped with a 10-foot boom, fitted with six TJ-VS 8002 nozzles spaced 20 inches apart at 3.6 mph. Fungicides were applied on July 24 at silk/blister (R1/R2) growth stage. Disease ratings were assessed on August 18 at dent (R5) growth stage. Gray leaf spot (GLS) was rated by visually assessing the percentage (0–100%) of symptomatic leaf area on ear leaf, and five plants were assessed per plot and averaged before analysis. The two center rows of each plot were harvested on October 16, and yields were adjusted to 15.5% moisture. All data were analyzed in SAS 9.4 (SAS Institute, Cary, NC). A generalized linear mixed model analysis of variance was performed using PROC GLIMMIX. Values are least squares means, and values with different letters are significantly different based on the least squares difference test (α = 0.05).

In 2023, weather conditions were not favorable for disease development. GLS was present in the trial but only reached low severity. No significant differences between treatments and nontreated control were detected for GLS severity, harvest moisture, test weight, and yield of corn.

TABLE 36. *Effect of Treatment on Foliar Disease and Yield of Corn*

TREATMENT AND RATE/ACRE[z]	GLS[y] %	HARVEST MOISTURE %	TEST WEIGHT LB/BU	YIELD[x] BU/ACRE
Nontreated control	0.16	16.4	58.0	163.7
Veltyma 3.34 SC 7.0 fl oz	0.02	16.0	58.6	160.9
Delaro Complete 458 SC 8.0 fl oz	0.00	15.8	58.3	156.0
Aproach Prima 2.34 SC 6.8 fl oz	0.05	16.5	58.2	164.3
Adastrio 4.0 SC 8.0 fl oz	0.05	16.2	58.3	155.5
Miravis Neo 2.5 EC 13.7 fl oz	0.02	15.9	58.5	164.2
Trivapro 2.21 SE 13.7 fl oz	0.01	15.9	58.5	159.5
Headline AMP 1.68 SC 10.0 fl oz	0.03	15.9	58.9	161.2
Proline 480 SC 5.7 fl oz	0.07	17.0	57.6	161.2
Quadris 2.08 SC 6.0 fl oz	0.08	16.0	58.5	147.9
Tilt 3.6 EG 4.0 fl oz	0.08	16.9	57.6	167.5
P-value[w]	*0.1312*	*0.6041*	*0.5041*	*0.6833*

[z] Fungicide treatments were applied on July 24 at silk/blister (R1/R2) growth stage.

[y] Gray leaf spot (GLS) was visually assessed as a percentage (0–100%) of leaf area on five plants in each plot and averaged before analysis on August 18 at dent (R5) growth stage.

[x] Yields were adjusted to 15.5% moisture and harvested on October 16.

[w] All data were analyzed in SAS 9.4 (SAS Institute, Cary, NC). A generalized linear mixed model analysis of variance was performed using PROC GLIMMIX. Values are least squares means, and values with different letters are significantly different based on the least squares difference test (α = 0.05).

EVALUATION OF FUNGICIDES FOR FOLIAR DISEASES ON SOYBEAN IN SOUTHWESTERN INDIANA, 2023 (SOY23-02.SWPAC)

S. Shim and D. E. P. Telenko, Department of Botany and Plant Pathology, Purdue University West Lafayette, IN 47907-2054

SOYBEAN (*GLYCINE MAX* P29A19E)

Frogeye leaf spot, *Cercospora sojina*
Septoria brown spot, *Septoria glycines*
Cercospora leaf blight, *Cercospora kikuchii*

A trial was established at the Southwest Purdue Agricultural Center (SWPAC) in Knox County, Indiana. The trial was a randomized complete block design with four replications. Plots were 10 feet wide and 30 feet long and consisted of four rows, and the two center rows were used for evaluation. The previous crop was corn. Standard practices for soybean production in Indiana were followed. Soybean cultivar P29A19E was planted in 30-inch row spacing at a rate of 135,000 seed/acre on April 11. All foliar fungicide applications were applied at 15 gal/acre and 40 psi using a Lee self-propelled sprayer equipped with a 10-foot boom, fitted with six TJ-VS 8002 nozzles spaced 20 inches apart. Fungicides were applied on July 14 at beginning pod (R3) growth stage. Disease ratings were assessed on September 18 at full seed (R6). Frogeye leaf spot (FLS), Septoria brown spot (SBS), and Cercospora leaf blight (CLB) were rated by visually assessing a percentage (0–100%) of symptomatic leaf area in the upper and lower canopies on August 18. The two center rows of each plot were harvested on September 22, and yields were adjusted to 13% moisture. All data were analyzed in SAS 9.4 (SAS Institute, Cary, NC). A generalized linear mixed model analysis of variance was performed using PROC GLIMMIX. Values are least squares means, and values with different letters are significantly different based on the least squares difference test (α = 0.05).

In 2023, weather conditions were not favorable for disease development. FLS, SBS, and CLB were present in the trial but only reached low levels. There was no significant difference between treatments for all diseases (Table 37). No significant differences were detected for harvest moisture, test weight, and yield of soybean.

TABLE 37. *Effect of Treatment on Foliar Disease Severity and Yield of Soybean*

TREATMENT, RATE/ACRE, AND TIMING[z]	FLS[y] %	SBS[y] %	CLB[y] %	HARVEST MOISTURE %	TEST WEIGHT LB/BU	YIELD[x] BU/ACRE
Nontreated control	0.4	0.5	2.5	12.8	55.9	75.6
Topguard EQ 4.29 SC 5.0 fl oz	0.0	0.4	0.8	12.6	56.4	73.9
Lucento 4.17 SC 5.0 fl oz	0.2	0.4	0.8	12.7	56.6	69.1
Trivapro 2.21 SE 13.7 fl oz	0.3	0.3	1.5	12.8	56.1	68.3
Quadris 2.08 SC 6.0 fl oz	0.3	0.6	1.8	12.9	56.2	73.7
Veltyma 3.34 SC 7.0 fl oz	0.2	0.5	0.9	12.9	55.7	76.9
Revytek 3.33 LC 8.0 fl oz	0.1	0.4	0.8	12.9	56.1	73.4
Echo 720 SE 36.0 fl oz + Folicur 3.6F 4.0 fl oz + Topsin 4.5 FL 4.5 fl oz	0.4	0.4	0.5	12.8	56.1	67.4
Delaro Complete 458 SC 8.0 fl oz	0.3	0.3	0.5	12.8	56.1	71.9
Miravis Neo 2.5 SE 13.7 fl oz	0.2	0.4	0.6	13.1	56.4	68.2
P-value[w]	0.0801	0.4579	0.1287	0.7655	0.5439	0.5025

[z] Fungicide treatments were applied on July 14 at beginning pod (R3) growth stage and contained a nonionic surfactant (Preference) at a rate of 0.25% v/v.

[y] Foliar disease severity was rated on a scale of 0–100% with disease symptoms on August 18. FLS = frogeye leaf spot, SBS = Septoria brown spot, CLB = Cercospora leaf blight.

[x] Yields were adjusted to 13% moisture and harvested on September 22.

[w] All data were analyzed in SAS 9.4 (SAS Institute, Cary, NC). A generalized linear mixed model analysis of variance was performed using PROC GLIMMIX. Values are least squares means, and values with different letters are significantly different based on the least squares difference test (α = 0.05).

FUNGICIDE EVALUATION FOR FOLIAR DISEASES IN SOYBEAN IN SOUTHWESTERN INDIANA, 2023 (SOY23-04.SWPAC)

E. A. Duncan, S. Shim, and D. E. P. Telenko, Department of Botany and Plant Pathology, Purdue University West Lafayette, IN 47907-2054

SOYBEAN (*GLYCINE MAX* P29A19E)

Frogeye leaf spot, *Cercospora sojina*
Septoria brown spot, *Septoria glycines*
Cercospora leaf blight, *Cercospora kikuchii*

A trial was established at the Southwest Purdue Agricultural Center (SWPAC) in Knox County, Indiana. The experiment was a randomized complete block design with four replications. Plots were 10 feet wide and 30 feet long and consisted of four rows, and the two center rows were used for evaluation. The previous crop was corn. Standard practices for soybean production in Indiana were followed. Soybean cultivar P29A19E was planted in 30-inch row spacing at a rate of 135,000 seed/acre on April 11. Fungicides were applied on July 14 and July 24 at beginning pod (R3) and beginning seed (R5) growth stages, respectively. All foliar fungicide applications were applied at 15 gal/acre and 40 psi using a Lee self-propelled sprayer equipped with a 10-foot boom, fitted with six TJ-VS 8002 nozzles spaced 20 inches apart. Frogeye leaf spot (FLS), Septoria brown spot (SBS), and Cercospora leaf blight (CLB) were rated for disease severity by visually assessing the percentage of symptomatic leaf area on August 18. The two center rows of each plot were harvested on September 22, and yields were adjusted to 13% moisture. All data were analyzed in SAS 9.4 (SAS Institute, Cary, NC). A generalized linear mixed model analysis of variance was performed using PROC GLIMMIX. Values are least squares means, and values with different letters are significantly different based on the least squares difference test (α = 0.05).

In 2023, weather conditions were unfavorable for disease development. FLS, SBS, and CLB were present in the trial but only reached low levels. There was no significant effect of fungicide treatment on FLS, CLB, and SBS severity (Table 38). There was no significant difference between fungicide treatments and the non-treated control for test weight and yield of soybean.

TABLE 38. *Effect of Treatment on Foliar Diseases and Yield of Soybean*

TREATMENT, RATE/ACRE, AND TIMING[z]	FLS[y] %	SBS[y] %	CLB[y] %	TEST WEIGHT LB/BU	YIELD[x] BU/ACRE
Nontreated control	0.18	0.2	1.0	56.2	77.7
Delaro Complete 458 SC 8.0 fl oz at R3	0.40	0.1	0.4	55.3	87.4
Lucento 4.17 SC 5.0 fl oz at R3	0.20	0.1	0.9	56.1	81.1
Trivapro 2.21 SE 13.7 fl oz at R3	0.20	0.1	0.8	56.0	85.5
Miravis Neo 2.5 SE 13.7 fl oz at R3	0.05	0.1	0.4	55.6	84.0
Revytek 3.33 LC 8.0 fl oz at R3	0.18	0.1	1.0	55.9	84.6
Delaro Complete 458 SC 8.0 fl oz at R5	0.05	0.1	0.4	56.3	79.9
Lucento 4.17 SC 5.0 fl oz at R5	0.08	0.1	1.1	56.0	83.3
Trivapro 2.21 SE 13.7 fl oz at R5	0.10	0.1	0.9	56.0	80.1
Miravis Neo 2.5 SE 13.7 fl oz at R5	0.05	0.1	0.8	55.5	86.9
Revytek 3.33 LC 8.0 fl oz at R5	0.03	0.1	1.1	56.0	83.7
Nontreated control	0.20	0.1	1.1	56.1	81.6
P-value[w]	*0.2640*	*0.4671*	*0.5900*	*0.3375*	*0.2709*

[z] Fungicides were applied on July 14 and July 24 at the at beginning pod (R3) and beginning seed (R5) growth stages. All treatments contained a nonionic surfactant (Preference) at a rate of 0.25% v/v.

[y] Foliar disease severity was rated by visually assessing the percentage of symptomatic leaf area in the upper and lower canopies on August 18. FLS = frogeye leaf spot, SBS = Septoria brown spot, CLB = Cercospora leaf blight.

[x] Yields were adjusted to 13% moisture and harvested on September 22.

[w] All data were analyzed in SAS 9.4 (SAS Institute, Cary, NC). A generalized linear mixed model analysis of variance was performed using PROC GLIMMIX. Values are least squares means, and values with different letters are significantly different based on the least squares difference test (α = 0.05).

EVALUATION OF FUNGICIDE EFFICACY FOR SCAB MANAGEMENT IN SOUTHWESTERN INDIANA, 2023 (WHT23-04.SWPAC)

E. A. Duncan, S. Shim, and D. E. P. Telenko, Department of Botany and Plant Pathology, Purdue University West Lafayette, IN 47907-2054

WHEAT (*TRITICUM AESTIVUM* P25R40)

Fusarium head blight, *Fusarium graminearum*

A trial was established at the Southwest Purdue Agricultural Center (SWPAC) in Knox County, Indiana. The experiment was a randomized complete block design with four replications. Plots were 7.5 feet wide and 20 feet long and consisted of 12 rows spaced 7.5 inches apart, and the center of each plot was used for evaluation. The previous crop was corn. Wheat cultivar P25R40 was drilled at 7.5-inch spacing on October 17, 2022. All fungicide applications were applied at 15 gal/acre and 40 psi using a CO_2 backpack sprayer equipped with a 10-foot boom, fitted with six TJ-VS 8002 nozzles spaced 20 inches apart and directed forward and backward at a 45-degree angle. Fungicides were applied on May 10 at the Feekes growth stage 10.5.1 and 5 days after on May 15. All plots were inoculated with a mixture of isolates of *Fusarium graminearum* endemic to Indiana on May 10 with a spore suspension (50,000 spores/ml) applied at 300 ml/plot with the CO_2 handheld sprayer. Disease ratings were assessed on May 30. Fusarium head blight (FHB) incidence was measured as the number of infected heads out of 60 plants in each plot and calculated as a percentage. FHB severity was rated by visually assessing the percentage (0–100%) of the infected heads. The FHB index was calculated as (% FHB incidence multiplied by average FHB severity)/100 per plot. The eight center rows of each plot were harvested with a Kincaid plot combine on June 21, and yields were adjusted to 13.5% moisture for comparison. A subsample of grain was taken from each plot and partitioned for DON (deoxynivalenol) analysis completed by the University of Minnesota DON testing lab and to determine Fusarium damaged kernels (FDK) by visually assessing the percentage (0–100%) of the infected heads. All data were analyzed in SAS 9.4 (SAS Institute, Cary, NC). A generalized linear mixed model analysis of variance was performed using PROC GLIMMIX. Values are least squares means, and values with different letters are significantly different based on the least squares difference test (α = 0.05).

In 2023, weather conditions were not favorable for FHB. A low level of FHB was detected in the trial. FHB incidence and index were significantly reduced by all fungicide applications except Caramba when compared to the nontreated control (Table 39). The concentration of DON was significantly reduced by all fungicide applications (Table 39). There were no significant differences between fungicide applications and nontreated control for FHB severity, percentage FDK, and yield.

TABLE 39. *Effect of Fungicide on Fusarium Head Blight (FHB), Deoxynivalenol (DON), Fusarium Damaged Kernels (FDK), and Yield of Wheat*

TREATMENT AND RATE/ACRE[z]	FHB DI[y]	FHB DS[x]	FHB INDEX[w]	FDK[v] %	DON[u] PPM	YIELD[t] BU/ ACRE
Nontreated control	10.4 a	7.0	0.7 a	8.0	0.29 a	71.1
Prosaro 421 SC 6.5 fl oz at 10.5.1	3.8 bc	1.2	0.1 b	7.0	0.10 b	72.9
Caramba 90 EC 13.5 fl oz at 10.5.1	7.5 ab	3.3	0.3 ab	6.6	0.05 bc	71.8
Miravis Ace 5.2 SC 13.7 fl oz at 10.5.1	2.1 c	4.1	0.3 b	10.5	0.05 bc	70.5
Prosaro Pro 400 SC 10.3 fl oz at 10.5.1	1.7 c	2.3	0.1 b	8.9	0.00 c	77.0
Sphaerex 300 EC 7.3 fl oz at 10.5.1	1.7 c	1.9	0.1 b	7.6	0.04 bc	71.0
Miravis Ace 5.2 SC 13.7 fl oz at 10.5.1 fb Prosaro Pro 400 SC 6.5 fl oz at 10.5.1 + 5 d	2.5 c	1.6	0.1 b	11.0	0.11 b	71.6
Miravis Ace 5.2 SC 13.7 fl oz at 10.5.1 fb Sphaerex 300 EC (BAS 84000F) 7.3 fl oz 10.5.1 + 5 d	4.6 bc	1.5	0.1 b	8.8	0.00 c	66.9
Miravis Ace 5.2 SC 13.7 fl oz at 10.5.1 fb Tebuconazole 3.6 SC 4.0 fl oz at 10.51 + 5 d	1.7 c	1.8	0.0 b	9.4	0.04 bc	71.0
P-value[s]	0.0093	0.3066	0.0379	0.2335	0.0001	0.7155

[z] Fungicide treatments were applied on May 10 and May 15 at the Feekes growth stage 10.5.1 and 10.5.1 + 5 days, respectively. All treatments contained a nonionic surfactant (Preference) at a rate of 0.125% v/v. All were plots inoculated with *Fusarium graminearum* spore suspension (50,000 spores/ml) after the treatment at Feekes 10.5.1. Spore suspension applied at 300 ml/plot. fb = followed by.

[y] FHB disease incidence (DI) was measured as the number of infected heads out of 60 plants in each plot and calculated as a percentage on May 30.

[x] FHB disease severity (DS) was rated by visually assessing the percentage of the infected head. FHB = Fusarium head blight on May 30.

[w] The FHB index was calculated as (FHB DI multiplied by average FHB DS)/100 per plot.

[v] Fusarium damaged kernels (FDK) was visually assessed as a percentage of Fusarium damaged heads.

[u] Analysis of the mycotoxin deoxynivalenol (DON) was completed by the University of Minnesota DON Testing Lab on August 17.

[t] Yields were adjusted to 13.5% moisture and harvested on June 21.

[s] All data were analyzed in SAS 9.4 (SAS Institute, Cary, NC). A generalized linear mixed model analysis of variance was performed using PROC GLIMMIX. Values are least squares means, and values with different letters are significantly different based on the least squares difference test ($\alpha = 0.05$).

EVALUATION OF FOLIAR FUNGICIDES AND CULTIVARS FOR SCAB MANAGEMENT IN SOUTHERN INDIANA, 2023 (WHT23-05.SWPAC)

E. A. Duncan, S. Shim, and D. E. P. Telenko, Department of Botany and Plant Pathology, Purdue University West Lafayette, IN 47907-2054

WHEAT (*TRITICUM AESTIVUM* P25R40 AND P25R61)

Fusarium head blight, *Fusarium graminearum*

A trial was established at the Southwest Purdue Agricultural Center (SWPAC) in Knox County, Indiana. The experiment was a strip-plot design with four replications. Plots were 7.5 feet wide and 20 feet long and consisted of 12 rows spaced 7.5 inches apart, and the center of each plot was used for evaluation. The previous crop was corn. On October 17, 2022, wheat cultivars P25R40 and P25R61 were drilled at 7.5-inch spacing. Fungicides were applied on May 10 at the Feekes growth stage 10.5.1. All fungicide applications were applied at 15 gal/acre and 40 psi using a CO_2 backpack sprayer equipped with a 10-foot boom, fitted with six TJ-VS 8002 nozzles spaced 20 inches apart and directed forward and backward at a 45-degree angle. Plots were inoculated with a mixture of isolates of *Fusarium graminearum* endemic to Indiana on May 10. Disease ratings were assessed on May 30. Fusarium head blight (FHB) incidence was measured as the number of infected heads out of 60 plants in each plot and calculated as a percentage (0–100%). FHB severity was rated by visually assessing the percentage (0–100%) of the infected head. The FHB index was calculated as (% FHB incidence multiplied by average FHB severity)/100 per plot. The eight center rows of each plot were harvested with a Kincaid 8XP combine on June 21, and yields were adjusted to 13.5% moisture. All data were analyzed in SAS 9.4 (SAS Institute, Cary, NC). A generalized linear mixed model analysis of variance was performed using PROC GLIMMIX. Values are least squares means, and values with different letters are significantly different based on the least squares difference test ($\alpha = 0.05$).

In 2023, weather conditions were not favorable for FHB. A low level of FHB was detected in the trial. FHB incidence was significantly reduced by Miravis Ace, Prosaro Pro, and Sphaerex compared to both the inoculated and noninoculated, nontreated controls (Table 40). FHB severity and FHB index were significantly higher in the noninoculated, nontreated control than any of the other treatments and the inoculated, nontreated control. The concentration of deoxynivalenol was significantly reduced by planting scab-resistant wheat cultivar P25R61 versus the scab-susceptible wheat cultivar P25R40. The concentration of deoxynivalenol was significantly reduced by the application of Prosaro and Sphaerex over both nontreated controls. There were no significant differences between cultivars for FHB incidence, FHB severity, FHB index, and yield of wheat. There was no significant difference between treatments for wheat yield.

TABLE 40. *Effect of Fungicide on Fusarium Head Blight (FHB), Deoxynivalenol (DON), Fusarium Damaged Kernels (FDK), and Yield of Wheat*

CULTIVAR OR TREATMENT AND RATE/ACRE[z]	FHB DI %[y]	FHB DS %[x]	FHB INDEX[w]	DON[v] PPM	YIELD[u] BU/ACRE
P25R40 (scab susceptible)	3.6	4.4	0.3	0.12 a	69.1
P25R61 (scab resistant)	2.6	3.6	0.2	0.03 b	66.4
Nontreated control, inoculated control	4.8 ab	5.6 b	0.4 ab	0.22 a	69.6
Nontreated, noninoculated control	6.3 a	11.7 a	1.0 a	0.12 b	64.9
Prosaro 421 SC 6.5 fl oz	2.3 bc	1.5 b	0.0 b	0.01 c	63.9
Miravis Ace 5.2 SC 13.7 fl oz	1.9 c	1.4 b	0.0 b	0.05 bc	68.7
Prosaro Pro 400 SC 10.3 fl oz	1.9 c	2.0 b	0.1 b	0.05 bc	71.2
Sphaerex 300 EC 7.3 fl oz	1.7 c	1.8 b	0.1 b	0.00 c	68.3
P-value *cultivar*[t]	*0.2117*	*0.6425*	*0.7875*	*0.0029*	*0.1660*
P-value *fungicide*	*0.0046*	*0.0049*	*0.0373*	*0.0003*	*0.2454*
P-value *cultivar*fungicide*	*0.5150*	*0.9917*	*0.8796*	*0.2103*	*0.1941*

[z] Fungicide treatments were applied on May 11 at Feekes growth stage 10.5.1. All treatments contained a nonionic surfactant (Preference) at a rate of 0.125% v/v. All plots were inoculated with *Fusarium graminearum* spore suspension (50,000 spores/ml) after the treatment at Feekes 10.5.1. Spore suspension were applied at 300 ml/plot with handheld sprayer on May 10.

[y] Fusarium head blight (FHB) incidence (DI) was measured as the number of infected heads out of 60 plants in each plot and calculated as a percentage (0–100%) on May 30.

[x] FHB severity (DS) was rated by visually assessing the percentage (0–100%) of the infected head on May 30.

[w] FHB index was calculated as (FHB incidence multiplied by average FHB severity)/100 per plot.

[v] Analysis of the mycotoxin deoxynivalenol (DON) was completed by the University of Minnesota DON Testing Lab.

[u] Yields were adjusted to 13.5% moisture and harvested on June 21

[t] All data were analyzed in SAS 9.4 (SAS Institute, Cary, NC). A generalized linear mixed model analysis of variance was performed using PROC GLIMMIX. Values are least squares means, and values with different letters are significantly different based on the least squares difference test (α = 0.05).

DAVIS PURDUE AGRICULTURAL CENTER (DPAC)

FIELD-SCALE EVALUATION OF DRONE VERSUS GROUND EQUIPMENT ON CORN DISEASES IN CENTRAL INDIANA, 2023 (COR23-08.DPAC)

M. S. Mizuno, S. Shim, and D. E. P. Telenko, Department of Botany and Plant Pathology, Purdue University West Lafayette, IN 47907-2054

CORN (*ZEA MAYS* P0574AM)

Tar spot, *Phyllachora maydis*
Gray leaf spot, *Cercospora zeae-maydis*

A trial was established at the Davis Purdue Agricultural Center (DPAC) in Randolph County, Indiana. The experiment was a randomized complete block design with three replications. Plots were 30 feet wide and 465 feet long and consisted of 12 rows, and the two center rows were used for evaluation. The previous crop was soybean. Standard practices for grain corn production in Indiana were followed. Corn hybrid P0574AMXT was planted in 30-inch row spacing at a rate of 32,000 seeds/acre on May 19. Veltyma 3.34 S 7.0 fl oz/acre was applied on August 1 at silk (R1) and August 23 at milk (R3) growth stages using two different applicators: a Case IH sprayer equipped with a 30-foot boom, fitted with 18 AIC110006 nozzles spaced 20 inches apart at 10 mph, and a DJI Agras T30 drone with spray pattern using 16 XRT TeeJet 11001VS nozzles spaced 20 inches apart at 47.0 mph, applied at 2 gal/acre, and at 11.6 mph, applied at 5 gal/acre. Disease ratings were assessed on September 20 at dent (R5) growth stage. Tar spot severity and gray leaf spot were visually assessed as a percentage (0–100%) of symptomatic leaf area on five plants per plot at three locations in each plot and averaged before analysis. Percent of canopy greenness was rated by visually assessing the percentage (0–100%) of canopy green on September 20 at dent (R5) growth stage. The trial was harvested on November 1, and yields were adjusted to 15.5% moisture. Data were averaged before analysis and subjected to mixed model analysis of variance using PROC GLIMMIX in SAS 9.4 (SAS Institute, Cary, NC). Values are least squares means, and values with different letters are significantly different based on the least squares difference test (α = 0.05).

In 2023, weather conditions were moderately favorable for disease. Tar spot and GLS were the most prominent disease in the trial and reached low severity. Tar spot severity was significantly reduced over the non-treated control by all application types and rates except drone at 2 GPA and 5 GPA applied at R3. There was no significant effect of treatment for GLS severity. There was no significant difference in treatments for percentage of canopy green, harvest moisture, and corn yield.

TABLE 41. *Effect of Different Application Type on Foliar Disease Severity, Canopy Greenness, and Yield of Corn*

APPLICATION EQUIPMENT, GPA, AND TIMING[z]	TAR SPOT[y] %	GLS[y] %	CANOPY GREEN[x] %	HARVEST MOISTURE %	YIELD[w] BU/ACRE
Nontreated control	0.53 a	0.2	90.5	25.0	194.3
DJI Agras T30 Drone, 2 GPA at R1	0.05 d	0.0	80.0	24.7	188.1
DJI Agras T30 Drone, 5 GPA at R1	0.28 b	0.0	80.9	24.6	188.5
Ground-rig, 20 GPA at R1	0.20 bc	0.0	88.9	24.4	192.3
DJI Agras T30 Drone, 2 GPA at R3	0.43 ab	0.0	90.1	24.7	184.9
DJI Agras T30 Drone, 5 GPA at R3	0.37 ab	0.1	79.8	25.4	187.4
Ground-rig, 20 GPA at R3	0.09 c	0.0	89.8	24.5	191.0
P-value[v]	0.0145	0.2466	0.6576	0.7094	0.9586

[z] Fungicide applications of Veltyma 3.34 S 7.0 fl oz/acre were made on August 1 at silk (R1) and August 23 at milk (R3) growth stages and contained a nonionic surfactant (Preference) at a rate of 0.25% v/v. GPA = gallons per acre.

[y] Foliar disease severity was visually assessed as a percentage (0–100%) of leaf area on five plants in each plot at the ear leaf on September 20 at dent (R5) growth stage. GLS = gray leaf spot.

[x] Canopy greenness was visually assessed as a percentage (0–100%) of canopy on September 20.

[w] Yields were adjusted to 15.5% moisture and harvested on November 1.

[v] All data were analyzed in SAS 9.4 (SAS Institute, Cary, NC). A generalized linear mixed model analysis of variance was performed using PROC GLIMMIX. Values are least squares means, and values with different letters are significantly different based on the least squares difference test (α = 0.05).

FIELD-SCALE EVALUATION OF DRONE VERSUS GROUND EQUIPMENT ON SOYBEAN DISEASES IN EAST CENTRAL, INDIANA (SOY23-07.DPAC)

M. S. Mizuno, S. Shim, and D. E. P. Telenko, Department of Botany and Plant Pathology, Purdue University West Lafayette, IN 47907-2054

SOYBEAN (*GLYCINE MAX* P29A19E)

Frogeye leaf spot, *Cercospora sojina*
Septoria brown spot, *Septoria glycines*

A trial was established at the Davis Purdue Agricultural Center (DPAC) in Randolph County, Indiana. The experiment was a randomized complete block design with three replications. Plots were 30 feet wide and 460 feet long and consisted of 24 rows, and the two center rows were used for evaluation. The previous crop was corn. Standard practices for soybean production in Indiana were followed. Soybean cultivar P29A19E was planted in 7.5-inch row spacing at a rate of 150,000 seeds/acre on May 17. Delaro Complete 458 SC 8.0 fl oz/acre was applied on August 1 at beginning pod (R3) and August 23 at beginning seed (R5) growth stages using two different applicators: a Case IH sprayer equipped with a 30-foot boom, fitted with 18 AIC110006 nozzles spaced 20 inches apart at 10 mph, and a DJI Agras T30 drone with spray pattern using 16 XRT Tee-Jet 11001VS nozzles spaced 20 inches apart at 47.0 mph, applied at 2 gal/acre, and at 11.6 mph, applied at 5 gal/acre. Disease ratings were assessed on September 7 at full seed (R6) growth stage. Frogeye leaf spot (FLS) was rated in the upper and lower canopies, and Septoria brown spot (SBS) was rated in the lower canopy. Severity of each disease was visually assessed as a percentage (0–100%) of symptomatic in three locations for each plot. All ratings were averaged in each plot before analysis. Soybean plots were harvested on October 12, and yields were adjusted to 13% moisture. All disease and yield data were analyzed using a mixed model analysis of variance, and means were separated using Fisher's least significant difference (α = 0.05).

In 2023, very little disease developed in plots. FLS and SBS were the most prominent diseases in the trial and reached low severity. There was no significant effect between application type and nontreated control for FLS upper and lower canopies and SBS on September 7 (Table 42). There was no difference between application type for soybean yield.

TABLE 42. *Effect of Different Application Type on Disease Severity in Soybean and Yield of Corn*

APPLICATION EQUIPMENT, GPA, AND TIMING[z]	FLS[Y] UPPER CANOPY %	FLS[Y] LOWER CANOPY %	SBS[Y] %	YIELD[x] BU/ACRE
Nontreated control	1.6	0.1	2.8	51.3
DJI Agras T30 Drone, 2.0 GPA at R3	1.1	0.2	1.1	53.7
DJI Agras T30 Drone, 5.0 GPA at R3	0.5	0.0	1.7	53.3
Ground-rig, 20.0 GPA at R3	0.4	0.0	2.6	56.3
DJI Agras T30 Drone, 2.0 GPA at R5	0.6	0.3	1.2	49.4
DJI Agras T30 Drone, 5.0 GPA at R5	0.6	0.1	1.8	51.6
Ground-rig, 20.0 GPA at R5	1.4	0.4	3.0	51.8
P-value[w]	*0.2438*	*0.3355*	*0.5328*	*0.3613*

[z] Fungicide applications of Delaro Completed 485 SC at 8.0 fl oz/acre were made on August 1 at beginning pod (R3) and on August 23 at beginning seed (R5) growth stages and contained a nonionic surfactant (Preference) at a rate of 0.25% v/v. GPA = gallons per acre.

[y] Foliar disease incidence was rated on a scale of 0–100% of plants with disease symptoms on September 7 at full seed (R6) growth stage. FLS = frogeye leaf spot in upper and lower canopies, SBS = Septoria brown spot in lower canopy.

[x] Yields were adjusted to 13% moisture and harvested on October 12.

[w] All data were analyzed in SAS 9.4 (SAS Institute, Cary, NC). A generalized linear mixed model analysis of variance was performed using PROC GLIMMIX. Values are least squares means, and values with different letters are significantly different based on the least squares difference test (α = 0.05).

NORTHEAST PURDUE AGRICULTURAL CENTER (NEPAC)

FIELD-SCALE EVALUATION OF DRONE VERSUS GROUND EQUIPMENT ON CORN DISEASES IN NORTHEAST INDIANA, 2023 (COR23-09.NEPAC)

M. S. Mizuno, S. Shim, and D. E. P. Telenko, Department of Botany and Plant Pathology, Purdue University West Lafayette, IN 47907-2054

CORN (*ZEA MAYS* P0574AM)

Tar spot, *Phyllachora maydis*
Northern corn leaf blight, *Exserohilum turcicum*

A trial was established at the Northeast Purdue Agricultural Center (NEPAC) in Whitley County, Indiana. The experiment was a randomized complete block design with six replications. Plots were 30 feet wide and 400 feet long and consisted of 12 rows, and the two center rows were used for evaluation. The previous crop was soybean. Standard practices for grain corn production in Indiana were followed. Corn hybrid P0574AM was planted in 30-inch row spacing at a rate of 32,000 seeds/acre on May 26. Veltyma 3.34 S 7.0 fl oz/acre was applied on August 8 at silk (R1) growth stage using two different applicators: a Case IH 2250 Patriot sprayer equipped with a 30-foot boom, fitted with 18 AITTJ60–11008VP nozzles spaced 20 inches apart, at 8 mph and applied at 15 gal/acre and 60 psi, and a DJI Agras T10 drone equipped with a 2.1-galkib spray tank with spray pattern using four TJ-VS 8002 nozzles spaced 20 inches apart and applied at 2.0 and 5.0 gal/acre and 40 psi. Disease rating was assessed on October 2 at dent (R5) growth stage. Tar spot severity and northern corn leaf blight (NCLB) were visually assessed as a percentage (0–100%) of symptomatic leaf area on five plants per plot at three locations in each plot and averaged before analysis. Percent of canopy greenness was rated by visually assessing the percentage (0–100%) of canopy green on October 2 at dent (R5) growth stage. The trial was harvested on November 2, and yields were adjusted to 15.5% moisture. Data were averaged before analysis and subjected to mixed model analysis of variance in SAS 9.4 (SAS Institute, Cary, NC). A generalized linear mixed model analysis of variance was performed using PROC GLIMMIX. Values are least squares means, and values with different letters are significantly different based on the least squares difference test ($\alpha = 0.05$).

In 2023, weather conditions were favorable for disease. Tar spot and NCLB were the most prominent diseases in the trial and reached low severity. All application types reduced tar spot severity over the non-treated control on October 2, with application using the DJI at 5 GPA resulting in the lowest level of tar spot (Table 43). There was no significant effect of application on NCLB severity on October 2. There was no significant difference in treatments for canopy greenness, harvest moisture, and corn yield.

TABLE 43. *Effect of Different Application Type on Foliar Disease Severity, Canopy Greenness, and Yield of Corn*

APPLICATION EQUIPMENT AND GPA[z]	TAR SPOT[y] %	NCLB[y] %	CANOPY GREEN[x] %	HARVEST MOISTURE %	YIELD[w] BU/ACRE
Nontreated control	1.4 a	1.0	68.6	32.3	190.3
DJI Agras T10 Drone, 2.0 GPA	0.9 b	0.4	63.3	32.3	191.5
DJI Agras T10 Drone, 5.0 GPA	0.7 c	0.0	69.2	33.2	189.8
Ground-rig, 20.0 GPA	0.9 b	0.2	72.2	32.6	195.8
P-value[v]	*0.0001*	*0.1907*	*0.4289*	*0.2200*	*0.6930*

[z] Fungicide treatment of Veltyma 3.34 S 7.0 fl oz/acre was applied on August 8 at silk (R1) growth stage using a ground-rig (15 GPA) and a drone (2 GPA and 5 GPA). All foliar treatments contained a nonionic surfactant (Preference) at a rate of 0.25% v/v. GPA = gallons per acre.

[y] Foliar disease severity was visually assessed as a percentage (0–100%) of leaf area on five plants in each plot at the ear leaf on October 2 at dent (R5) growth stage. NCLB = northern corn leaf blight.

[x] Canopy greenness was visually assessed as a percentage (0–100%) of canopy green on October 2.

[w] Yields were adjusted to 15.5% moisture and harvested on November 2.

[v] All data were analyzed in SAS 9.4 (SAS Institute, Cary, NC). A generalized linear mixed model analysis of variance was performed using PROC GLIMMIX. Values are least squares means, and values with different letters are significantly different based on the least squares difference test (α = 0.05).

FIELD-SCALE EVALUATION OF DRONE VERSUS GROUND EQUIPMENT ON SOYBEAN DISEASES IN NORTHEAST INDIANA, 2023 (SOY23-09.NEPAC)

M. S. Mizuno, S. Shim, and D. E. P. Telenko, Department of Botany and Plant Pathology, Purdue University West Lafayette, IN 47907-2054

SOYBEAN (*GLYCINE MAX* P29A19E)

Frogeye leaf spot, *Cercospora sojina*
Septoria brown spot, *Septoria glycines*

A trial was established at the Northeast Purdue Agricultural Center (NEPAC) in Whitley County, Indiana. The experiment was a randomized complete block design with six replications. Plots were 30 feet wide and 360 feet long and consisted of 48 rows, and the two center rows were used for evaluation. The previous crop was corn. Standard practices for soybean production in Indiana were followed. Soybean cultivar P29A19E was planted in 7.5-inch row spacing at a rate of 200,000 seeds/acre on May 17. Delaro Complete 458 SC at 8.0 fl oz/acre was applied on August 8 at beginning pod (R3) growth stage using two different applicators: a Case IH 2250 Patriot sprayer equipped with a 30-foot boom, fitted with 18 AITTJ60–11008VP nozzles spaced 20 inches apart, at 8 mph and applied at 15.0 gal/acre and 60 psi, and a DJI Agras T10 drone equipped with a 2.1-gallon spray tank with spray pattern using four TJ-VS 8002 nozzles spaced 20 inches apart and applied at 2.0 and 5.0 gal/acre and 40 psi. Disease ratings were assessed on September 8 at full seed (R6) growth stage. Frogeye leaf spot (FLS) was rated in the upper and lower canopies, and Septoria brown spot (SBS) was rated in the lower canopy. Severity of each disease was visually assessed as a percentage (0–100%) of symptomatic leaf area in three locations for each plot. All ratings were averaged in each plot before analysis. Soybean plots were harvested on October 3, and yields were adjusted to 13% moisture. All disease and yield data were analyzed using a mixed model analysis of variance, and means were separated using Fisher's least significant difference ($\alpha = 0.05$).

In 2023 weather conditions were not favorable for diseases, and very little disease developed in plots. FLS and SBS were the most prominent disease in the trial and reached low severity. There was no significant effect between application type and the nontreated control for FLS in the upper and lower canopies and SBS severity on September 7 (Table 44). There was no difference between application type for soybean yield.

TABLE 44. *Effect of Different Application Type on Disease Severity and Yield of Soybean*

APPLICATION EQUIPMENT AND GPA[z]	FLS[y] UPPER CANOPY %	FLS[y] LOWER CANOPY %	SBS[y] %	BU/ACRE YIELD[x]
Nontreated control	0.4	0.9	1.6	48.1
DJI Agras T10, 2.0 GPA	0.5	1.0	2.2	48.1
DJI Agras T10, 5.0 GPA	0.4	0.6	1.9	46.4
Ground-rig, 20.0 GPA	0.3	0.4	1.8	47.2
P-value[w]	0.8613	0.1846	0.8383	0.4011

[z] Fungicide applications of Delaro Completed 458 SC at 8.0 fl oz/acre were made on August 8 at beginning pod (R3) growth stage and contained a nonionic surfactant (Preference) at a rate of 0.25% v/v.

[y] Foliar disease incidence was rated on a scale of 0–100% of plants with disease symptoms on September 7 at full seed (R6) growth stage. FLS = frogeye leaf spot in upper and lower canopies, SBS = Septoria brown spot in lower canopy.

[x] Yields were adjusted to 13% moisture and harvested on October 3.

[w] All data were analyzed in SAS 9.4 (SAS Institute, Cary, NC). A generalized linear mixed model analysis of variance was performed using PROC GLIMMIX. Values are least squares means, and values with different letters are significantly different based on the least squares difference test (α = 0.05).

SOUTHEAST PURDUE AGRICULTURAL CENTER (SEPAC)

FIELD-SCALE EVALUATION OF DRONE VERSUS GROUND EQUIPMENT ON CORN DISEASES IN SOUTHEASTERN INDIANA, 2023 (COR23-10.SEPAC)

M. S. Mizuno, S. Shim, and D. E. P. Telenko, Department of Botany and Plant Pathology, Purdue University West Lafayette, IN 47907-2054

CORN (*ZEA MAYS* P0574AM)

Tar spot, *Phyllachora maydis*
Gray leaf spot, *Cercospora zeae-maydis*

A trial was established at the Southeast Purdue Agricultural Center (SEPAC) in Jennings County, Indiana. The experiment was a randomized complete block design with four replications. Plots were 30 feet wide and 1,794 feet long and consisted of 12 rows, and the two center rows were used for evaluation. The previous crop was soybean. Standard practices for grain corn production in Indiana were followed. Corn hybrid P1077AM was planted in 30-inch row spacing at a rate of 30,378 seeds/acre on May 18. Veltyma 3.34 S 7.0 fl oz/acre was applied on July 27 at silk (R1) growth stage using two different applicators: an Apache AS720 sprayer equipped with a 30-foot boom, fitted with six TTJ60–11005 nozzles spaced 15 inches apart at 12 mph and applied at 20.0 gal/acre and 60 psi, and a DJI Agras T30 drone with spray pattern using 16 XRTTeeJet 11001VS nozzles spaced 20 inches apart at 47.0 mph, applied at 2.0 gal/acre and at 11.6 mph applied at 5.0 gal/acre. Disease ratings were assessed on September 14 at dent (R5) growth stage. Tar spot severity and gray leaf spot (GLS) were visually assessed as a percentage (0–100%) of symptomatic leaf area at ear leaf on five plants per plot at three locations in each plot and averaged before analysis. Percent of canopy greenness was rated by visually assessing the percentage (0–100%) of canopy green on September 5 at dent (R5) growth stage. The trial was harvested on October 31, and yields were adjusted to 15.5% moisture. Data were averaged before analysis and subjected to mixed model analysis of variance in SAS 9.4 (SAS Institute, Cary, NC). A generalized linear mixed model analysis of variance was performed using PROC GLIMMIX. Values are least squares means, and values with different letters are significantly different based on the least squares difference test (α = 0.05).

In 2023, weather conditions were not favorable for diseases. Tar spot and GLS were the most prominent diseases in the trial and reached low severity. There was no significant effect of application type on tar spot stroma severity on September 14 (Table 45). All applications significantly reduced GLS severity over the nontreated control on September 14, but there was no difference between application type. There was no significant difference in treatments for percentage of canopy green, harvest moisture, and harvest moisture. Veltyma sprayed with the drone at 2 GPA and ground-rig significantly increased yield over the nontreated control.

TABLE 45. *Effect of Different Application Types on Foliar Disease Severity, Canopy Greenness, and Yield of Corn*

APPLICATION EQUIPMENT AND RATE/ACRE[z]	TAR SPOT[y] %	GLS[y] %	CANOPY GREEN[x] %	HARVEST MOISTURE %	YIELD[w] BU/ACRE
Nontreated control	1.1	1.5 a	69.6	18.5	228.1 b
DJI Agras T30 Drone, 2.0 GPA	0.0	0.4 b	77.5	19.1	235.4 a
DJI Agras T30 Drone, 5.0 GPA	0.1	0.1 b	70.8	18.7	229.9 b
Ground-rig, 20.0 GPA	0.4	0.7 b	69.2	19.2	237.0 a
P-value[v]	*0.3987*	*0.0473*	*0.2704*	*0.1111*	*0.0136*

[z] Fungicide treatment with Veltyma 3.34 S at 7.0 fl oz/acre was applied on July at silk (R1) growth stage using a ground-rig and a drone with 2 GPA and 5 GPA. All foliar treatments contained a nonionic surfactant (Preference) at a rate of 0.25% v/v using a ground rig and Maaytx 1 oz/acre. GPA = gallons per acre.

[y] Tar spot stroma severity and gray leaf spot (GLS) severity were visually assessed as a percentage (0–100%) of leaf area on five plants in each plot at the ear leaf on September 14 at dent (R5) growth stage.

[x] Canopy greenness was visually assessed as a percentage (0–100%) of canopy green on September 5.

[w] Yields were adjusted to 15.5% moisture and harvested on October 31.

[v] All data were analyzed in SAS 9.4 (SAS Institute, Cary, NC). A generalized linear mixed model analysis of variance was performed using PROC GLIMMIX. Values are least squares means, and values with different letters are significantly different based on the least squares difference test (α = 0.05).

FIELD-SCALE EVALUATION OF DRONE VERSUS GROUND EQUIPMENT ON SOYBEAN IN SOUTHEASTERN INDIANA, 2023 (SOY23-08.SEPAC)

M. S. Mizuno, S. Shim, and D. E. P. Telenko, Department of Botany and Plant Pathology, Purdue University West Lafayette, IN 47907-2054

SOYBEAN (*GLYCINE MAX* P29A19E)

Frogeye leaf spot, *Cercospora sojina*
Septoria brown spot, *Septoria glycines*
Cercospora leaf blight, *Cercospora kikuchii*

A trial was established at the Southeast Purdue Agricultural Center (SEPAC) in Jennings County, Indiana. The experiment was a randomized complete block design with five replications. Plots were 30 feet wide and 650 feet long and consisted of 24 rows, and the two center rows were used for evaluation. The previous crop was corn. Standard practices for soybean production in Indiana were followed. Soybean cultivar P29A19E was drilled in 15-inch row spacing at a rate of 129,000 seeds/acre on May 11. Delaro complete 458 SC 8.0 fl oz/acre was applied on July 24 and July 27 at beginning pod (R3) and on August 17 at beginning seed (R5) growth stages using two different applicators an Apache AS720 sprayer equipped with a 30-foot boom, fitted with six TTJ60–11005 nozzles spaced 15 inches apart at 12 mph, applied at 20 gal/acre and 60 psi, and a DJI Agras T30 drone with spray pattern using 16 XRTTeeJet 11001VS nozzles spaced 20 inches apart at 47.0 mph, applied at 2 gal/acre and at 11.6 mph applied at 5 gal/acre. Disease rating was assessed on September 14 at beginning maturity (R7) growth stage. Frogeye leaf spot (FLS) was rated in the upper and lower canopies, Septoria brown spot (SBS) was rated in the lower canopy, and Cercospora leaf blight (CLB) was rated in the upper canopy. Severity of each disease was visually assessed as a percentage (0–100%) of symptomatic in three locations in each plot. Soybean plots were harvested on September 25, and yields were adjusted to 13% moisture. All data were analyzed in SAS 9.4 (SAS Institute, Cary, NC). A generalized linear mixed model analysis of variance was performed using PROC GLIMMIX. Values are least squares means, and values with different letters are significantly different based on the least squares difference test (α = 0.05).

In 2023 weather conditions were not favorable for diseases, and very little disease developed in plots. FLS, SBS, and CLB were the most prominent diseases in the trial and reached low severity. No significant differences were detected for the applications compared to the nontreated control except that the application of Delaro Complete with the ground rig at R3 had higher severity of FLS in the upper canopy on September 14 (Table 45). There was no significant effect between application type and the nontreated control for FLS in the lower canopy, DM, and CLB severity. Septoria brown spot (SBS) was reduced by all applications and timings of compared to the nontreated control, but there was no difference between application type or timing. There were no significant differences between application type or timing for soybean yield.

TABLE 45. *Effect of Different Application Type on Disease Severity and Yield of Soybean*

APPLICATION EQUIPMENT, GPA, AND TIMING[z]	FLS[y] UPPER CANOPY %	FLS[y] LOWER CANOPY %	SBS[y] %	CLB[y] %	YIELD[x] BU/ACRE
Nontreated control	1.1 b	0.1	46.3 a	0.2	62.9
DJI Agras T30 Drone, 2.0 GPA at R3	0.6 b	0.0	6.1 b	0.0	65.4
DJI Agras T30 Drone, 5.0 GPA at R3	0.7 b	0.1	5.0 b	0.4	67.3
Ground-rig, 20.0 GPA at R3	1.8 a	0.0	7.5 b	0.5	65.7
DJI Agras T30 Drone, 2.0 GPA at R5	0.8 b	0.1	12.0 b	0.3	64.6
DJI Agras T30 Drone, 5.0 GPA at R5	0.7 b	0.0	3.9 b	0.2	63.7
Ground-rig, 20 GPA	1.0 b	0.0	12.9 b	0.2	62.1
P-value[w]	0.0082	0.5364	0.0001	0.5906	0.0518

[z] Fungicide treatment of Delaro Complete 458 SC at 8.0 fl oz/acre was applied on July 27 at silk (R1) growth stage using a ground rig and a drone with 2 GPA and 5 GPA. All foliar treatments contained a nonionic surfactant (Preference) at a rate of 0.25% v/v using the ground rig and Maaytx 1 oz/acre. GPA = gallons per acre.

[y] Foliar disease incidence was rated on a scale of 0–100% of canopy with disease symptoms on September 14 at beginning maturity (R7) growth stage. FLS = frogeye leaf spot in upper and lower canopies, SBS = Septoria brown spot in lower canopy, CLB = Cercospora leaf blight.

[x] Yields were adjusted to 13% moisture and harvested on September 25.

[w] All data were analyzed in SAS 9.4 (SAS Institute, Cary, NC). A generalized linear mixed model analysis of variance was performed using PROC GLIMMIX. Values are least squares means, and values with different letters are significantly different based on the least squares difference test ($\alpha = 0.05$).

COMPARISON OF PLANTING DATES AND SEED TREATMENTS ON SOYBEAN IN SOUTHEASTERN INDIANA. (SOY23-12.SEPAC)

I. L. Miranda, J. R. Wahlman, and D. E. P. Telenko, Department of Botany and Plant Pathology, Purdue University West Lafayette, IN 47907-2054

SOYBEAN (*GLYCINE MAX* P29A19E)

A trial was established at Southeast Purdue Agricultural Center (SEPAC) in Butlerville, Indiana. The experiment was a randomized complete block design with three replications. Plots were 15 feet wide and 900 feet long and consisted of six rows, and the two center rows were used for evaluation. The previous crop was corn. Standard practices for soybean production in Indiana were followed. Soybean seeds were planted in 30-inch row spacing at a rate of 130,000 seeds/acre. Treatments were a factorial arrangement of four planting dates by four seed treatments. Soybeans were planted on April 12 (planting date 1), April 26 (planting date 2), May 10 (planting date 3), and May 25 (planting date 4). Stand counts were assessed at cotyledons expanded/first-node (VC/V1) growth stage for each planting date. Ten roots were sampled from the outer rows of each plot and rated for root rot severity on a scale of 0–100% and averaged before analysis. Root dry weight was calculated from the 10 sampled roots. Each plot was harvested on September 30, and yields were adjusted to 13% moisture. All data were analyzed in SAS 9.4 (SAS Institute, Cary, NC). A generalized linear mixed model analysis of variance was performed using PROC GLIMMIX. Values are least squares means, and values with different letters are significantly different based on a least squares difference test (α = 0.05).

In 2023 very little disease developed in plots, although low levels of Septoria brown spot (SBS) were detected. There was no significant interaction between planting date and seed treatment for root rot, root dry weight, and yield; therefore, main effects are presented. There was a significant interaction between planting date and seed treatment for soybean stand count (data not shown). Looking at only the main effects, soybean stand was the highest at planting on May 10 as compared to all the other planting dates. Also, treatment with CruiserMaxx APX with thiamethoxam and thiamethoxam alone resulted on the greatest stand counts compared to the nontreated and CruiserMaxx without thiamethoxam (Table 46). Root rot severity was significantly higher at planting on April 26 as compared to all the other planting dates. There were no significant differences between seed treatments for root rot. There were no significant differences between planting dates and seed treatments for root weight. Soybean yield was significantly reduced at planting on May 4 and 25 as compared to earlier planting dates. No significant differences were detected between seed treatments and soybean yield.

TABLE 46. *Effect of Planting Dates and Seed Treatments on Stand count, Root Rot, Root Weight and Soybean Yield*

PLANTING DATES AND SEED TREATMENTS[z]	STAND COUNT #/ ACRE	ROOT ROT %[y]	ROOT DRY WEIGHT G[x]	YIELD BU/ ACRE[w]
Planting Date				
Planting date 1 (April 12)	54,250 d	6.9 ab	26.2	71.7 a
Planting date 2 (April 26)	69,642 b	9.2 a	24.4	68.2 b
Planting date 3 (May 10)	88,336 a	4.0 bc	27.4	67.3 b
Planting date 4 (May 25)	62,636 c	0.4 c	29.0	58.0 c
Seed Treatment				
Nontreated control	61,783 b	5.7	25.2	65.4
CruiserMaxx APX + thiamethoxam	75,776 a	6.1	27.9	65.9
Thiamethoxam	71,620 a	4.1	28.2	66.6
CruiserMaxx APX no thiamethoxam	65,685 b	4.8	25.6	67.5
P-value *planting date*[v]	0.0001	0.0011	0.1898	0.0001
P-value *seed treatment*	0.0001	0.7544	0.3755	0.1931
P-value *planting date*seed treatment*	0.0001	0.9363	0.7300	0.4865

[z] Seed treatments were applied prior to planting at 10 g AI/100 kg seed.

[y] Root rot was visually assessed as a percentage (0–100%) of dark discoloration on roots on September 12.

[x] Root dry weight was calculated from the 10 dried root samples in grams (g).

[w] Yields were adjusted to 13% moisture and harvested on September 30.

[v] All data were analyzed in SAS 9.4 (SAS Institute, Cary, NC). A generalized linear mixed model analysis of variance was performed using PROC GLIMMIX. Values are least squares means, and values with different letters are significantly different based on the least squares difference test (α = 0.05).

APPENDIX: WEATHER DATA

TABLE 47. *Average Monthly Weather Conditions at the Purdue Agronomy Center for Research and Education (ACRE), the Pinney Purdue Agricultural Center (PPAC), the Southwest Purdue Agricultural Center (SWPAC), the Davis Purdue Agricultural Center (DPAC), the Northeast Purdue Agricultural Center (NEPAC), and the Southeast Purdue Agricultural Center (SEPAC) in Indiana, 2023[z]*

| | ACRE | | | | PPAC | | | | SWPAC | | | |
| | TEMPERATURE | | | TOTAL | TEMPERATURE | | | TOTAL | TEMPERATURE | | | TOTAL |
MONTHS	AVE[y] °F	MAX[y] °F	MIN[y] °F	PRECIPITATION[x] (INCHES)	AVE[y] °F	MAX[y] °F	MIN[y] °F	PRECIPITATION[x] (INCHES)	AVE[y] °F	MAX[y] °F	MIN[y] °F	PRECIPITATION[x] (INCHES)
January	34.34	38.94	29.45	0.71	31.46	37.15	26.65	1.68	39.11	46.53	32.98	3.84
February	37.27	47.93	26.99	3.86	32.95	42.27	24.04	3.52	42.78	53.83	32.68	2.76
March	40.00	49.69	31.27	4.50	35.83	44.19	27.96	4.90	44.72	54.38	35.73	6.60
April	52.95	66.02	40.07	1.26	49.14	61.12	36.16	1.32	56.13	68.15	45.09	2.89
May	65.22	77.72	52.93	2.72	60.29	73.05	47.50	2.37	66.46	78.70	54.97	2.05
June	70.22	83.59	56.86	1.03	66.62	78.10	55.39	3.39	73.02	84.91	62.14	4.52
July	73.89	85.08	62.96	6.04	70.24	81.67	58.77	4.65	77.46	87.89	68.52	2.98
August	71.65	82.53	61.42	4.34	68.48	79.32	58.12	3.93	74.92	85.60	66.07	4.98
September	66.10	80.43	53.26	0.33	63.38	75.25	52.16	1.97	69.81	84.19	58.71	0.45
October	55.57	66.53	45.56	2.81	52.71	62.28	43.62	6.17	58.40	69.77	48.99	2.25
November	42.93	54.61	31.87	0.49	39.78	50.32	29.66	0.68	47.59	59.59	37.04	0.67
December	39.50	46.92	33.25	2.48	37.35	43.07	31.83	2.58	42.18	50.68	35.28	2.56
Annual	*54.14*	*65.00*	*43.82*	*30.57*	*50.69*	*60.65*	*40.99*	*37.16*	*57.71*	*68.69*	*48.18*	*36.55*

MONTHS	DPAC				NEPAC				SEPAC			
	TEMPERATURE			TOTAL PRECIPITATION[x]	TEMPERATURE			TOTAL PRECIPITATION[x]	TEMPERATURE			TOTAL PRECIPITATION[x]
	AVE[y] °F	MAX[y] °F	MIN[y] °F	AVE[y] (INCHES)	AVE[y] °F	MAX[y] °F	MIN[y] °F	MAX[y] (INCHES)	AVE[y] °F	MAX[y] °F	MIN[y] °F	(INCHES)
January	34.70	41.27	28.10	3.26	33.16	38.87	27.76	4.22	38.78	46.35	30.72	1.81
February	37.16	48.84	26.13	2.60	34.89	44.12	26.26	3.02	42.53	54.27	30.82	4.00
March	39.41	49.45	30.67	4.49	37.52	46.45	29.27	8.04	43.43	54.66	33.00	5.27
April	50.98	63.22	38.49	2.43	50.36	61.83	38.75	4.26	54.11	67.92	41.50	2.29
May	60.99	73.42	47.97	2.55	61.09	72.74	48.84	1.11	63.80	76.68	51.05	3.14
June	67.75	79.47	52.37	3.33	68.41	79.96	56.78	2.62	69.81	83.02	57.40	3.40
July	73.17	84.24	62.59	3.90	72.68	82.73	63.05	7.05	74.72	86.52	64.93	5.97
August	69.75	80.53	58.93	3.20	69.23	79.55	59.36	3.26	72.69	85.76	61.94	3.16
September	64.86	78.56	52.33	0.99	65.02	76.69	54.55	0.28	67.32	83.52	54.69	1.18
October	54.50	65.50	44.27	3.24	53.80	63.08	45.37	3.15	56.47	68.78	45.29	3.69
November	42.03	54.49	29.83	0.98	41.26	51.06	31.56	0.93	45.70	59.22	33.72	0.72
December	39.05	46.84	32.05	1.17	38.75	44.80	33.15	2.23	40.95	50.55	32.19	2.21
Annual	*52.86*	*63.82*	*41.98*	*32.14*	*52.18*	*61.82*	*42.89*	*40.17*	*55.86*	*68.10*	*44.77*	*36.84*

[z] Data courtesy of Indiana State Climate Office, Beth Hall, Jonathan Weaver, and Austin Pearson, https://ag.purdue.edu/indiana-state-climate/. Taken from Purdue Mesonet stations.

[y] Average minimum and maximum temperatures for each month.

[x] Total precipitation for each month.

ABOUT THE AUTHORS

DARCY E. P. TELENKO is an associate professor and Extension plant pathologist in the Department of Botany and Plant Pathology at Purdue University. Her interdisciplinary research and Extension program are involved in studying the biology and management of soilborne and foliar pathogens of agronomic crops. Telenko is a native of western New York and received her PhD at North Carolina State University. She has published more than sixty peer-reviewed manuscripts and two hundred Extension publications. Telenko was awarded the 2024 Leadership Award from the Purdue University Cooperative Extension Specialist Association.

SUJOUNG SHIM is a research associate in the Department of Botany and Plant Pathology at Purdue University. Her research involves designing, conducting, analyzing, and reporting on a variety of research projects. She has a BS in pharmaceutical science and an MS in public health, both from Purdue University. Shim has served as a coauthor on more than ten peer-reviewed publications and twenty-five peer-reviewed technical reports.

www.ingramcontent.com/pod-product-compliance
Lightning Source LLC
Chambersburg PA
CBHW082105210326
41599CB00033B/6597

* 9 7 8 1 6 2 6 7 1 2 4 9 2 *